Scientific Secrets of Nature

W0115132

Springer Nature More Media App

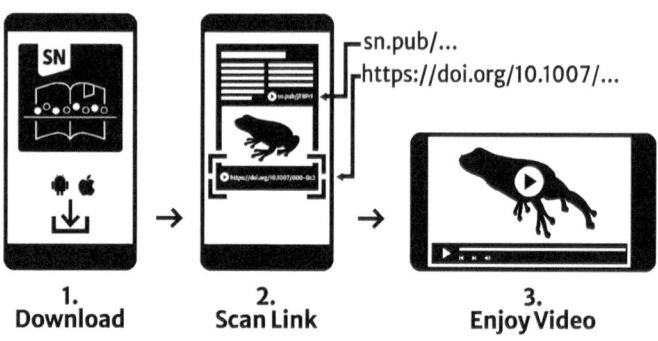

sn.pub/...
https://doi.org/10.1007/...

1.
Download

2.
Scan Link

3.
Enjoy Video

Support: customerservice@springernature.com

Andreas Korn-Müller

Scientific Secrets of Nature

Indoor & Outdoor Experiments for the Whole Family

 Springer

Andreas Korn-Müller
Dresden, Germany

ISBN 978-3-662-69574-6   ISBN 978-3-662-69575-3   (eBook)
https://doi.org/10.1007/978-3-662-69575-3

Translation from the German language edition: "Der Natur auf der Spur" by
Andreas Korn-Müller, © Springer Verlag GmbH, DE 2023. Published by Springer
Berlin Heidelberg. All Rights Reserved.

This book is a translation of the original German edition "Der Natur auf der Spur"
by Andreas Korn-Müller, published by Springer-Verlag GmbH, DE in 2023. The
translation was done with the help of an artificial intelligence machine translation
tool. A subsequent human revision was done primarily in terms of content, so that
the book will read stylistically differently from a conventional translation. Springer
Nature works continuously to further the development of tools for the production
of books and on the related technologies to support the authors.

This Springer imprint is published by the registered company Springer-Verlag
GmbH, DE, part of Springer Nature.
The registered company address is: Heidelberger Platz 3, 14197 Berlin, Germany

If disposing of this product, please recycle the paper.

Preface

Dear readers,
Dear students,
Dear children and young researchers,

Nature is full of surprises! If you look closer, you will discover spectacular physical phenomena, fascinating biochemical reactions, and exciting biological structures. With the help of the experiments and explanations, you will hopefully get closer to nature in a playful way and with a lot of fun, and thus get a bit closer. In addition, it becomes clear that the three major sciences of biology, chemistry, and physics are always interconnected and essentially form a symbiosis, with which we can better understand nature and the environment—in the forest, in the park, in the garden, on the meadow, by the stream, by the pond, by the sea. For all four seasons, I have tried to offer suitable experiments, which are marked with a corresponding seasonal logo. This is what they look like:

Spring:

Summer:

Autumn:

Winter:

These logos indicate in which season the experiment can be (best) conducted. Often, several logos are visible at the same time, as many of the experiments cover a wide range of months. If a season is particularly well suited, then the corresponding logo is double-bordered.

I have divided my book into six chapters, which include both indoor and outdoor experiments.

1. Amazing Plants (indoor experiments): Carnivorous plants, such as sundew and Venus flytrap, can be "fed" with gummy bears. They are mercilessly digested.

Kohlrabi leaves display fantastic lotus effects with honey, ink, and graphite powder. Videos are available for viewing.

2. In the garden, on the meadow, terrace, or balcony (indoor and outdoor experiments): The green leaf (chlorophyll), the autumn coloring, and photosynthesis are discussed, followed by experiments with chlorophyll solutions. With their red glow, you can trace photosynthesis. The greenhouse effect in the bottle is about climate change. Grass gets hot, a bush becomes a light organ with a laser pointer, and the cress grows as you want.

3. In the forest, in the park, on the hike (outdoors): I hope the enthusiasm for lichens also overcomes you, because with them you can estimate the current air quality quite well. No matter where you live, with the lichen grid you can determine the air quality directly on site. Yellow lichens glow beautifully in UV light. Two methods for measuring tree height are described and the secret of the opened pine cones is also revealed.

4. At the pond, puddle, or lake (outdoors and indoors): With the "laser drop method" you can make invisible microorganisms, cells, and plankton visible, without a microscope! It's incredible what kind of creatures are bustling in the waters. Spectacular videos are available for viewing. In a soap bubble, you can very nicely observe the light reflection and the refraction of a laser pointer beam. But if you send the laser beam through a small ice floe, a real disco effect is created.

5. The dark side of nature (outdoors and indoors): With a UV flashlight, you embark on a glowing trail search in the dark, as numerous natural objects glow (fluoresce): fruit, vegetables, a creepy pepper, moss, mushrooms, duckweed, lichens, woodlice, minerals but also trash in the bushes can be found with it.

6. At the beach (outdoors): On the North Sea, I had the pleasure of experiencing the marine luminescence close—an unforgettable experience. My children were thrilled. If you couldn't enjoy the blue glow on the North or Baltic Sea, no problem, with a UV flashlight you can create a wonderful glow on the beach: algae, seaweed, crab shells, mussels, and jellyfish fluoresce red, orange, yellow, and blue—a real color spectacle at night. During the day, children can create sand avalanches or—if they feel like it—collect chalk fossils, chicken gods, and amber on the Baltic Sea. How far can you let your gaze wander to the horizon? You can read the answer here.

Now I wish you all a lot of joy, fun, and success in experimenting with nature, and I hope that you will learn a lot of new and interesting things through my compilation. Perhaps my book will also become your constant travel companion when you head out into nature or to the sea. Be on the trail of nature!

Your Andreas Korn-Müller—"Magic Andy"

Andreas Korn-Müller

Acknowledgment

I am very grateful to my son Melvin Müller for creating the graphics and tables!

My thanks go to Mr. Dipl.-Phys. Reinhard Fink, Mr. Dr. rer. nat. Till Biskup, and Mr. Harald Steinhofer for the stimulating scientific exchange and valuable advice.

I am very grateful to Mrs. Elisabeth Link for the enlightening and informative lichen excursion in the Black Forest.

Contents

About the Author

Andreas Korn-Müller He studied chemistry in Tübingen and received his doctorate in 1994 at the MPI for Biochemistry in Martinsried. After two years of post-doc research in the HIV high-security laboratory of LMU Munich, he has been working as a freelancer in the field of science communication since 1997. In addition to various exhibitions at museums, the multiple award-winning laureate has developed eight different science shows, which he successfully performs worldwide at science festivals for young and old under the stage name "Magic Andy". So far, he has written four (children's) non-fiction books as well as numerous contributions for professional journals.

1

Amazing Plants

Abstract Take a gummy bear and a carnivorous plant and soon the poor gummy bear is done for. It is digested, softened, and dissolved, skin and all. Within a few days, you can watch the digestive juices of sundew or Venus flytrap at work—creepy! Snap shut—gummy bear mush. The lotus effect of kohlrabi leaves is more civilized. Honey finds no hold on the leaf surface and ink is absorbed as soon as a drop of water comes into play. Graphite powder can "stick" to any surface—but not to the kohlrabi leaf. In this chapter, you can also watch amazing videos about the lotus effect. Nature offers with autumn leaves and sulfur lichens further masters of the lotus effect that want to be discovered. Rainy weather? Ideal! For indoors, you can

---

Supplementary Information The electronic version of this chapter contains additional material, which can be accessed via the following link https://doi.org/10.1007/978-3-662-69575-3_1. The videos can be played by clicking on the DOI link in the legend of a corresponding figure, or by scanning this link with the SN More Media App.

prepare a pH indicator from radish peels—color change games included. All experiments are easy to replicate.

1.1   Carnivorous Plants

Who doesn't know them, the so-called "carnivorous" plants, from documentaries or as gigantic mutated eating monsters in absurd horror films or even from animated films like "Ice Age". Carnivores are plants that can catch and digest, i.e., eat up small animals—especially insects. However, they do not do this with teeth, chewing, saliva, and stomach acid, but with a special liquid. This liquid contains digestive enzymes that dissolve the shell, tissue, and organs of the prey. Insects consist, among other things, of their chitin shell (long sugar chains) and protein (long amino acid chains), with the carnivores mainly interested in the protein. The digestive enzymes can be imagined as molecular hedge clippers and chainsaws, which attack the captured prey millions of times and cut it into invisibly small protein pieces. It takes many days for a captured animal to be completely digested. Poor creature! The protein pieces are finally transported into the plant and serve as nutrients for the growth of the carnivore. The main reason for the pronounced desire for "meat" is primarily nutrient-poor soil. There are essentially three types of carnivores [1, 2]: adhesive traps, snap traps, and pitfall traps. The fact that the digestive juice of carnivorous plants is actually capable of dissolving proteins and carbohydrates (sugars) can be verified with an impressive experiment—with the native sundew and the exotic Venus flytrap. Since commercially available gummy bears also contain protein, they are suitable as "food" for the carnivores (Fig. 1.1).

Fig. 1.1 Do carnivores actually hunt gummy bears? Graphic: Melvin Müller

1.1.1 Experiment: The Sundew—Plant Sucks Gummy Bears

You will need

- Cape Sundew plant (garden market, internet, approx. 5–9 €)
- Some gummy bears
- Magnifying glass
- Patience (3–9 days)

How it works

For this experiment, it is best to get the Cape Sundew *(Drosera capensis).* It has long adhesive tentacles, is easy to care for, and is available in garden markets or on the internet. Place two or three dark-colored gummy bears with their backs on several of the tentacle-like trapping arms of the sundew. Why on the back? Because the large, smooth back offers more "attack surface" than the front. The

Fig. 1.2 Gummy bear sticks to the adhesive tentacles of a sundew plant *(Drosera capensis)*

tentacles of the sundew are tipped with tiny sticky droplets, to which the gummy bears adhere (Fig. 1.2). They are also referred to as adhesive tentacles.

Make sure that the gummy bears lie straight and cannot slip. Why dark-colored gummy bears? So that the "feeding traces" can be seen better. After 3 days: Take a gummy bear off the trapping arm with your fingers and look at the back with a magnifying glass. You can already see the first small "feeding traces" as tiny notches and holes—like small knife cuts—on the gummy bear (Fig. 1.3a). After 5 to 6 days, you can already see larger holes and slots on the surface of the gummy bear. After 9 days, you will see clear holes and "craters" (Fig. 1.3b). Look at the traces under the magnifying glass! Pretty intense. Like acid, the sundew's digestive juice has eaten into the gummy bear.

What's the science behind it?

Sundew is a small, carnivorous plant that grows on all continents except Antarctica—especially in moor and swamp areas [1, 3]. The leaves of the sundew are equipped

Fig. 1.3 Digestion traces on gummy bears by sundew. **a** after 3 days (the circles mark the "craters"). **b** after 9 days

with numerous tentacles (trapping arms). Each tentacle secretes a sticky, sweet, tough, and decomposing fluid. So, the tentacle droplets contain everything a insectivore needs: sweeteners, adhesives, solvents (enzymes). An insect attracted by the sweet scent sticks to a trapping arm, which rolls up to enclose the prey. Neighboring tentacles notice the catch and also bend towards the defenseless insect and digest it collectively [2, 3]. The insects usually suffocate due to the secretion penetrating their respiratory openings (tracheae). Because gummy bears also consist of 6.9% protein (gelatin), the gelatin is attacked and dissolved by the enzymes. Gelatin is mainly composed of collagen, a long protein made up of thousands of amino acids [4].

1.1.2 Experiment: The Venus Flytrap: Snap Shut—Fly Dead (or Gummy Bear Mash)

You will need

- Venus flytrap plant (garden center, internet, approx. 6–9 €)
- Some gummy bears
- Scissors
- Brush or toothpick

How it works

For this experiment, get a Venus flytrap *(Dionaea muscipula)*. Cut a gummy bear into small pieces and place one of these "snacks" in an open trap. You can also stick a halved bear inside. To make the trap snap shut, you have to trick the plant into thinking that a live prey is in the trap. To do this, you must touch the tiny, red sensory hairs on the inside of the trap leaf at least twice within 30 s with a brush or toothpick, only then will the trap snap shut. There are usually three of these sensory hairs on each trap leaf half. The gummy bear should be completely enveloped. If the trap leaf opens again over the course of hours, you must make it snap shut again by artificially stimulating it with a brush/toothpick (Fig. 1.4).

After 2–4 days, the pieces are already mostly softened, decomposed, dissolved, and partially liquefied. The liquid gummy bear mush sometimes even drips out of the closed trap leaf (Fig. 1.5).

Fig. 1.4 Gummy bear enclosed by the trap of a Venus flytrap

Fig. 1.5 Partially digested gummy bear in the Venus flytrap after about four days

After about 4–6 days, you can cut off and open the "fed" trap leaf along with the gummy bear and look at the contents. Figure 1.6 shows the viscous slime of a digested gummy bear.

What's the science behind it?

The Venus flytrap *(Dionaea muscipula)* is a carnivorous plant that grows exclusively in bog areas on the east coast of the USA [2, 5]. Within 100 milliseconds, which is

Fig. 1.6 Liquefied gummy bear after about a week in the trap

a tenth of a second, the leaf traps snap shut and are thus world record holders for fast movements in the plant kingdom [6]. The insides of the trap leaves are studded with numerous glands, which secrete a digestive juice with numerous digestive enzymes after five or continuous touches of the sensory hairs. Proteins and carbohydrates are broken down and dissolved in liquid. So also the gummy bear, which consists of 6.9% protein (gelatin) and 77.4% carbohydrates, of which 45.6% is sugar. Gelatin mainly contains collagen, a long protein made up of thousands of amino acids [4].

By the way, you can also "feed" carnivorous plants with real meat or cheese. The meat is attacked, but it stinks terribly after a few days due to decay by bacteria.

Background

Nasty Trap Setters

The most famous, spectacular insect-catching plant in the world is probably the **Venus flytrap** *(Dionaea muscipula)*. Its leaf traps snap shut at an impressive speed of up to 0.1 m/s [7]. Once caught, there is no escape. Although plants do not have nerve pathways like humans or other mammals, the Venus flytrap uses a similar basic principle, namely that of electrochemical signal transmission [7]. Every slightest touch, every minimal bending of the sensory hairs leads to a release of calcium ions (Ca^{2+}) in the sensory cells at the base of the hair, which act as signal messengers. Corresponding receptors transmit the increasing number of Ca^{2+} ions to all cells over the entire trap leaf, like a stone thrown into water, whose "signal" spreads as concentric waves over the entire water surface. When the signals reach the leaf edges and the trap bristles, they cause the trap leaf to close. Conclusion: Mechanical stimulus triggers the electrochemical closing of the trap. Researchers have recently discovered that Venus flytraps can be anaesthetised with diethyl ether ("ether"), just like humans [7]. This well-known, centuries-old anaesthetic blocks the receptors of signal transmission and thus "paralyses" the leaf movement. The Venus flytrap can no longer close its trap leaves, no matter how many insects are crawling around in them and bending the sensory hairs endlessly. Sensational!

Also, **pitcher plants** *(Nepenthes)* have developed a nasty trapping method. Their pitchers are built like small wells: with a slippery, downwardly curved edge and filled with digestive juice deep down. The edges of the pitchers are sweet—but also slippery. Attracted insects crawl around on them, slip, lose their grip and—plump—fall into the deadly liquid. Very efficient. Even small animals like mice and frogs meet their Waterloo there. The largest specimens in the world are native to Southeast Asia and can grow up to 50 cm long [1].

The digestive fluid of pitcher plants reads like a wild mixture from an alchemical chemistry lab and consists, among other things, of proteases, peroxidases, esterases, phosphatases, ribonucleases and chitinases. The ending "-ase" always means in biochemistry that an enzyme is at work and often brings with it a breaking, dividing or splitting property. Examples: Proteases divide proteins, ribonucleases split genetic material and chitinases chop up chitin.

But there are also many enzymes that combine and build up substances, transfer electrons and even entire molecule groups. Examples: Ligases, transferases, polymerases [8].

1.2 Sparkling, Spotless Leaves: The Lotus Effect

There are plants whose leaves always look absolutely clean. The leaves of such super clean plants have a special coating on their leaf surface. Nothing sticks or stays on it. No dust, no dirt, not even bird droppings. The world champion in being spotless is the Asian lotus flower, a water plant similar to the water lily [9]. But also the leaves of the nasturtium and our native kohlrabi are real "cleaners". Figure 1.7 shows some kohlrabi leaves in a vase, with which you can perform amazing experiments.

1.2.1 Experiment: Honey that Doesn't Stick

You will need

- Some fresh kohlrabi leaves (supermarket, vegetable market, own garden)
- Water
- Liquid honey (in squeeze bottle)

Fig. 1.7 Kohlrabi leaves in a vase with water

How it works

For the experiments, the kohlrabi leaves should be fresh, firm, not wilted, and undamaged. Tip: Often you can find loose kohlrabi leaves at the vegetable stand or in the green waste of the supermarket, which you can take for free. If the leaves look "limp" and "flabby"—no problem: cut off the leaf stem fresh and put the leaf in a glass or vase with water, then they last longer and do not wilt (Fig. 1.7). After a short time, the leaves firm up and the experiments can begin.

1. Pour some water on a kohlrabi leaf! The water immediately beads off. The leaf stays dry.
2. Put a blob of honey on a leaf! Tilt the leaf! The honey balls up, flows slowly off and simply rolls off the leaf as a ball (Fig. 1.8). Almost nothing sticks to the leaf. Incredible!

Fig. 1.8   **a** Blob of honey on kohlrabi leaf. **b** Close-up

How the drop of honey effortlessly rolls off the kohlrabi leaf, you can retrieve in a short video via the URL in Fig. 1.9.

Try this with leaves from roses, primroses, oaks, beeches, maples, or chestnuts… Here the honey sticks and leaves a long "slime trail". A video is also available for this experiment. To watch the video, simply scan the URL in Fig. 1.10.

What's the science behind it?
What causes this reduced wettability of the leaves? If you look at a lotus leaf under an electron microscope, the surface of the leaf looks like a field with many small trees or

Fig. 1.9   The video shows a drop of honey rolling off a kohlrabi leaf. Music: Calm background for video by Ivy music (pixabay) URL: ▶https://doi.org/10.1007/000-a65

Fig. 1.10 The video shows a drop of honey on a primrose leaf. Music: Calm background for video by Ivy music (pixabay) URL: ▸https://doi.org/10.1007/000-a63

hills, the so-called papillae, which are about as large or tiny as bacteria. (Fig. 1.11a and b). You can imagine this leaf layer like the surface of an egg carton (Fig. 1.11c).

On and around the papillae is a wax layer, which can be seen in Fig. 1.11b as a blue-gray coating. Wax is absolutely water-repellent, the chemist says: hydrophobic. You know this from candles. A candle does not dissolve under water. Water beads off wax. The numerous "mountains and valleys" also cause a high roughness. This reduces the contact between water droplets and leaf surface to a minimum and the adhesion forces drop massively, almost to zero [9]. You can imagine it as if you are walking on tiptoes over a hot floor plate to have as little contact as possible. Both properties—roughness and wax—ensure the superhydrophobic properties of the leaves. Ultimately, the surface tension (cohesive forces) of the water causes the droplet to curve into a sphere. Figure 1.12a shows water droplets on a lotus leaf and in Fig. 1.12b you see the analogy model: The egg carton represents the leaf surface, the ball symbolizes a water droplet or a solid particle; a dirt particle, dirt, dust, paint or or or …

The surface of a kohlrabi leaf looks completely different, but causes the same superhydrophobic effect. Under

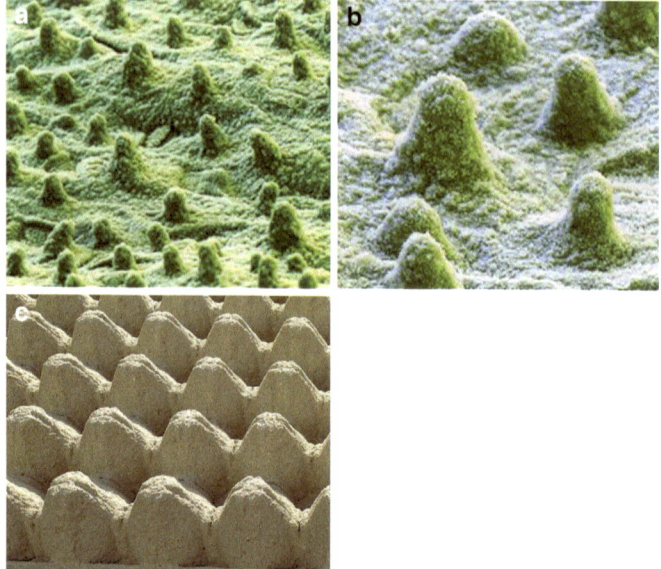

Fig. 1.11 **a** Scanning electron microscopic image of the leaf surface of a lotus leaf at 1500x and **b** at 3900x magnification. Size of the papillae: approx. 2–5 μm. The blue-gray coating is a wax layer. **c** Egg carton as a vivid model for the leaf surface of a lotus leaf. *Source*: Photos a and b from Science Photo/Eye of Science, No. 12526315 and No. 12526319

the electron microscope, you can see a "tangled", three-dimensional network of bizarre wax crystals [9]. The tips of these crystals create a minimal contact surface and thus minimal adhesion while simultaneously repelling water through the wax. Roughness and wax also ensure here that neither dust nor dirt find a hold on the leaf and are simply washed away by (rain) water. Even honey is repelled by the wax layer and cannot stick to the small hills or crystal tips. It simply beads off. A perfectly non-wettable surface perfected over the course of evolution.

Fig. 1.12 **a** Water droplets on an always clean lotus leaf. The hydrophobic layer causes the formation of water beads. **b** Ball on an egg carton. The contact between the ball surface and the bumps of the egg carton is minimal, so the ball can roll very easily over the egg carton. *Source*: Photo a from adobe stock No. 24362204

Background

Adhesion
Attractive forces between two different materials/substances, e.g. water and glass (water droplet runs along the edge of the glass when poured out slowly), adhesive on paper, licked finger for picking up crumbs, scraps of paper, etc.

Cohesion
Attractive forces between the same substances, e.g. water molecules in water, which quasi hold hands and build a kind of network (water, ice).

hydrophobic
Water-repellent/water-insoluble.

hydrophilic
Water-loving/water-soluble.

1.2.2 Experiment: Smooth or Rough—That is the Question!

You will need

- Sheet of paper
- Coffee filter or blotting paper
- Water

How it works

Drip a drop of water onto a sheet of paper and onto a filter paper (coffee filter, blotting paper) with your finger. While the drop on the filter paper immediately "disappears" and is absorbed, the drop on the regular paper remains relatively long and spreads a little in width. The drop shining in the light is clearly visible.

What's the science behind it?

Paper is made of cellulose, a long-chain molecule made up of thousands of glucose units. Cellulose can be classified as rather hydrophilic, or water-loving. Wetting with water is rather moderately successful on the hydrophilic, smooth paper surface. The drop remains flat on the surface and is only slowly absorbed. The filter paper, when viewed microscopically, is nothing more than roughened paper: as if the surface had been worked with sandpaper. This roughness drastically increases the paper's wettability, so that the water drop is immediately absorbed and largely dries up. Figure 1.13 illustrates this. The wettability of a surface is expressed by the contact angle α, which in turn

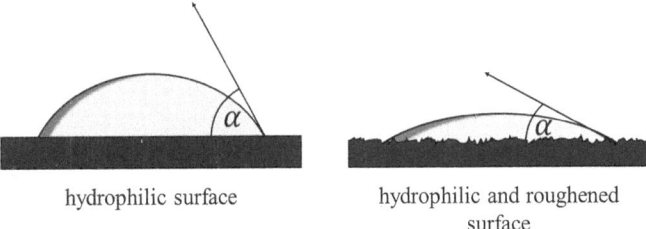

hydrophilic surface   hydrophilic and roughened
                                      surface

Fig. 1.13 Water drop on hydrophilic surface. Left: Paper (smooth). Right: Filter paper (roughened). The flat contact angle α indicates good wettability. (After [10]). Graphic: Melvin Müller

results from the surface tensions. The rule is: the smaller the contact angle, the better the wettability. In nature, the two extremes—contact angle α = 0° or 180°—do not occur. This would mean that the liquid lies flat on the surface (α = 0°) or touches the surface tangentially at only one point (α = 180°) [10].

A smooth and hydrophobic, i.e., water-repellent layer, such as a wax layer, is very difficult to wet with water. The water remains hemispherical on the wax, beads off, and is not absorbed. If this already water-repellent wax layer is additionally roughened, it mutates into superhydrophobic behavior: water or honey drops completely round off and bead off the surface without residue (Fig. 1.14).

The behavior of dirt on the rough, non-wettable leaf surfaces is also astonishing. If there is dirt on the leaves, the rain will wash away the dirt particles in the form of water drops. But different phenomena come into play depending on whether the dirt is water-soluble or water-repellent.

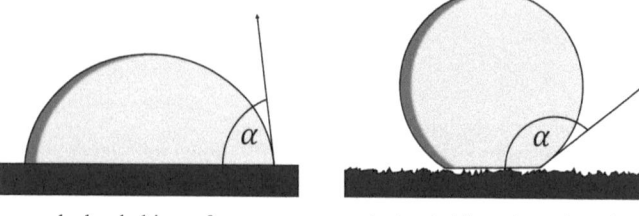

hydrophobic surface   hydrophobic and roughened surface

Fig. 1.14 Water drop on hydrophobic surface. Left: Wax layer (smooth). Right: Roughened wax layer. The large contact angle α indicates poor wettability. (After [10]). Graphic: Melvin Müller

1.2.3 Experiment: The Dirt Must Go

You will need

- 2–3 Fresh kohlrabi leaves (supermarket, vegetable market)
- Blue ink (e.g., from ink cartridge, fountain pen)
- Graphite powder (hardware store) or cocoa powder
- Spray bottle (flower sprayer)
- Water

How it works

First, apply 2–3 drops of blue ink to a kohlrabi leaf. Let it dry overnight. Now put a drop of water on the leaf and let the drop roll over the dried blue color. The blue ink dye is immediately detached from the leaf and absorbed into the water droplet because the dye is easily water-soluble.

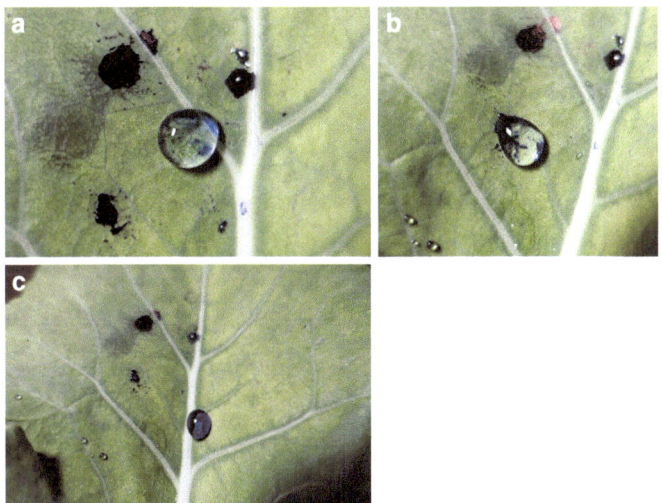

Fig. 1.15 **a** Water droplet on kohlrabi leaf with dried ink. **b** The water-soluble blue ink is absorbed by the water droplet. **c** The water droplet turns completely blue

In a matter of seconds, the water droplet is dyed deep blue (Fig. 1.15). A video can be accessed with Fig. 1.16.

You can also spray the leaf a few times with a sprayer. This easily removes the blue "dirt" and it collects as a spherical "puddle" in the center of the leaf if you hold the leaf slightly bent in your hand. After completing this experiment, you can simply pour the blue spherical droplets into the kitchen sink and wash off the leaf.

So far so clear. But what happens with water*insoluble* dirt? For this, one sprinkles graphite or coal powder or cocoa powder on a kohlrabi leaf. You can easily make coal powder yourself: Simply rub a piece of charcoal over fine sandpaper and tap the black dust onto the leaf. Now add one or more drops of water to the leaf. One would actually assume that the water-repellent, i.e., hydrophobic dirt would much rather stick to the also hydrophobic

Fig. 1.16 The video shows a water droplet absorbing blue ink on a kohlrabi leaf. Music: The beat of Nature by Olexy (pixabay) URL: ▸https://doi.org/10.1007/000-a64

wax layer of the leaf. But far from it. Instead, the dirt particles immediately adhere to the water droplet, and they do so until the entire spherical surface of the droplet is completely covered with particles, as shown in Fig. 1.17. In this process, no particles get *into* the water droplet, but exclusively onto the *surface*. When using graphite powder, the spherical surface is therefore covered with a silver-metallic shiny layer (Fig. 1.17). I think it looks great. Scan the URL in Figs. 1.18 and 1.19 and watch two impressive videos on this.

Spray the leaf with a water atomizer! Numerous small silver-gray beads form, their surface completely covered with graphite (Fig. 1.20). Watch the accompanying video, which can be accessed under Fig. 1.21. The "graphite drop" behaves almost like mercury!

What's the science behind it?
The blue ink dye is *water-soluble* (hydrophilic) and is therefore readily and easily absorbed by the water droplet. The droplet turns blue accordingly. The hydrophilic dye loves water and therefore immediately combines with it, leaving no residue of dye on the leaf. Self-cleaning made easy!

Fig. 1.17 **a** Graphite powder and water droplet on kohlrabi leaf. **b** The hydrophobic graphite powder adheres to the surface of the water droplet. **c** Almost completely covered droplet. **d** Droplet completely covered with graphite powder. Around the droplet, you can see the "cleaned" track

Fig. 1.18 The video shows the wetting of a water droplet with graphite powder on a kohlrabi leaf. Music: Mountain Path by Magnetic Trailer (pixabay) URL: ►https://doi.org/10.1007/000-a62

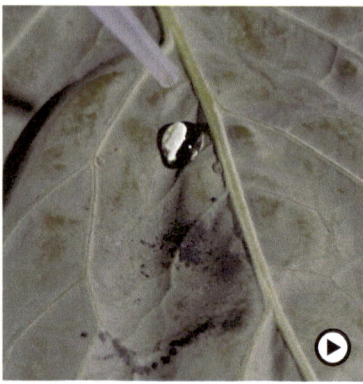

Fig. 1.19 The video shows the wetting of a water droplet with graphite powder on a kohlrabi leaf. Music: Mountain Path by Magnetic Trailer (pixabay) URL: ▶https://doi.org/10.1007/000-a66

Fig. 1.20 Graphite powder on a kohlrabi leaf sprayed with water. Numerous small silver-gray beads form

The situation is more complicated with *water-repellent* (hydrophobic) dirt, such as graphite, coal, cocoa, or pepper. Although these particles do not like water at all, the particles accumulate on the entire surface of the sphere until no space is left. A "trail" or "collection path" around the droplet is clearly visible on the leaf (Fig. 1.22).

However, no powder is absorbed into the droplet. The explanation is relatively simple: Even the hydrophobic

Fig. 1.21 The video shows numerous water droplets with graphite powder on a kohlrabi leaf, and how a large drop incorporates smaller drops. Music: Acoustic vibe by RomanSenykMusic (pixabay) URL: ▸https://doi.org/10.1007/000-a67

Fig. 1.22 Only the surface of the water droplet is coated with graphite. Inside is clear water. Around the droplet, you can see the clean "courtyard"

dirt particles only have a very small contact on the fine wax crystals and thus also exert only a very small adhesion force. The particles are practically not grabbed and held with a whole hand but only touched with the fingertips and thus hardly held. When water now flows over these particles held at the "wax fingertips", they adhere more strongly to the water. This means that the adhesion force between water and hydrophobic particle is stronger than the adhesive force between particle and the wax crystals [11]. This is indeed a remarkable, phenomenal natural effect.

1.2.4 Experiment: Mirror, Mirror on the Leaf

You will need

- Fresh kohlrabi leaf (supermarket, vegetable market)
- Glass of water

How it works

If you submerge a kohlrabi leaf in water, the leaf surface appears silvery and shiny—a kind of mirror effect (Fig. 1.23).

Similar effects are known from the water spider and from carbon dioxide gas bubbles in mineral water or sparkling wine. If you pull the leaf out of the water again, it is bone dry. Magic? No, science!

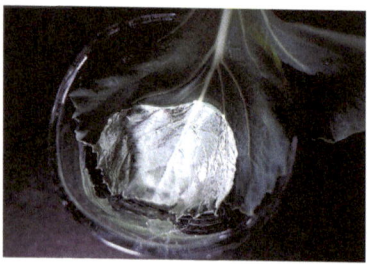

Fig. 1.23 A submerged kohlrabi leaf appears silvery and shiny

What's the science behind it?

The water-repellent (hydrophobic) wax layer of the leaf prefers to bond with air rather than water. As soon as you submerge the leaf in water, it drags a thin layer of air beneath the water surface. Thus, the leaf remains dry underwater because it is protected by an air envelope. The mirror effect or the silvery shine is created by (total) reflection of light at the interface between air and water.

1.2.5 Experiment: Destroyed "Lotus Effect"

You will need

- Fresh kohlrabi leaf (supermarket, vegetable market)
- Kitchen paper
- Water
- Graphite or cocoa powder
- Honey (from squeeze bottle)

How it works

Rub a kohlrabi leaf vigorously between your fingers with a piece of kitchen paper! Now the "lotus effect" no longer works or only very poorly. The "lotus effect" is gone and water wets the leaf like a normal leaf. Even honey sticks and leaves a trail like a slimy snail. Graphite or cocoa powder can no longer be absorbed with water.

What's the science behind it?

The fine, water-repellent wax crystals on the leaf surface are destroyed by mechanical forces. This can already be seen with the naked eye: The leaf no longer looks matte, but it shines and appears greener.

Background

The Lotus Effect

The German botanist Prof. Wilhelm Barthlott (*1946) was the first to research the self-cleaning of lotus plant leaves from 1997 [2, 11]. Therefore, he referred to this self-cleaning as the "lotus effect". Barthlott's discovery led to astonishing technical applications: self-cleaning glass panes, tarps and sails, never wet swimming suits, dirt-repellent clothing and building surfaces.

In addition to the lotus plant and kohlrabi, other plant leaves show the "lotus effect": including nasturtium, white cabbage, reed and columbine. But no leaf can do it as efficiently as the lotus plant.

1.2.6 Lotus Effect on Autumn Leaves in Drizzly Weather

Take a walk in the park or forest during the beautiful drizzly weather of late autumn/early winter, in November or December! Pay attention to the wet leaves on the ground, especially the oak leaves. If it's raining, you can look forward to a small sensation. Brown, dead leaves of the pedunculate oak surprise with their surface, on which water droplets lie like pearls next to each other. Figure 1.24a shows one of many oak leaves that I found on a rainy day in December in a park.

The hydrophobic surface of the underside of the leaf exhibits a "lotus effect" and allows water droplets to form

Fig. 1.24 **a** Raindrops on the underside of an autumn oak leaf. **b** Raindrops on different autumn leaves (maple, beech, linden, birch), each on the underside of the leaf. Only the oak leaf shows a "lotus effect"

into spheres. Other leaves, such as those from linden, maple, or chestnut, simply get wet and do not show any water-repellent properties (Fig. 1.24b). Perhaps the stomata on the underside of the leaf are protected from dust and dirt by this "lotus effect". Nature is always full of surprises.

Some lichens also have water-repellent surfaces, such as the sulfur lichens. In Fig. 1.25, you can clearly see the spherical water droplets on the lichen surface [12].

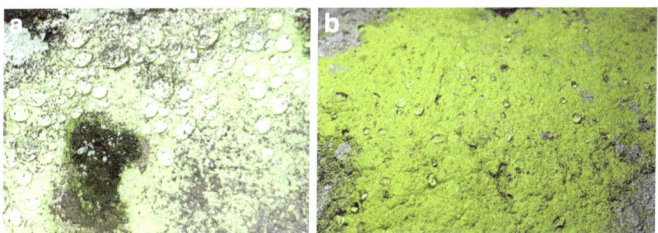

Fig. 1.25 "Lotus effect": Raindrops on **a** a yellow fruit sulfur lichen (yellow-green), **b** a rock sulfur lichen

1.3 Change Your Color!

Red cabbage juice is widely known as an acid-base indicator and can be found in almost every chemistry textbook. But radishes are also very suitable as an indicator *pro domo*. For the "holistic" approach, you (with your children) can easily grow radishes in a raised bed, garden, or flower box. Radishes are quite undemanding plants that grow quickly and can even be grown several times a year. If the waiting takes too long: market or supermarket.

1.3.1 Experiment: Indicator Solution from Radishes

You will need

- 6–7 Radishes
- Water
- Small pot and stove
- Measuring cup
- Small PET bottle (500 mL) with screw cap
- 3 Narrow glasses
- Lemon or vinegar
- Laundry detergent

How it works

Put the peels of six to seven radishes together with about 200 mL of water in a small pot and bring it to a boil. After a few minutes, remove the pot from the stove and let the broth cool down. The peels look pale pink to white, the water has turned purple. The hot water has thus extracted the dyes from the peels. Now you can filter the dye solution through a coffee filter into a small PET bottle. Then divide the purple solution evenly into two narrow glasses (Fig. 1.26a). Test as an acid indicator: Add some lemon juice or vinegar to the left glass. Swirl around. The color changes to red (Fig. 1.26b). Test as a base indicator: Add some laundry detergent to the right glass and swirl around. After dissolving the powder, the solution turns yellow (Fig. 1.26b).

Note: The radish solution only lasts a few days, then it becomes cloudy and starts to smell terribly. So please consume quickly or prepare fresh again and again.

What's the science behind it?

The peel of the radishes contains the purple dye pelargonin and belongs to the class of substances called anthocyanins [13]. The name is reminiscent of pelargoniums, those red hanging geraniums that you often see blooming on balconies. Like most anthocyanins, pelargonin is pH-sensitive. In a neutral solution, it shows a purple color

Fig. 1.26 **a** Radish dye extracted by boiling. **b** Left: Color change to red when adding lemon juice or vinegar. Right: Color change to yellow when adding laundry detergent

(Fig. 1.26a). In an alkaline environment, it is colored yellow, in an acidic environment its color changes to red (Fig. 1.26b). The cheerful "change your color" is based on a structural change due to H^+- or OH^--ions. Acids provide H^+-ions and bases (alkalis) contribute OH^--ions. The attachment of these charged particles causes a rearrangement of the electrons in the anthocyanin molecule, which ultimately shifts the absorption behavior towards light.

By the way, the sharpness of radishes is caused by a completely different substance: mustard oil, an isothiocyanate, a compound of nitrogen, carbon, and sulfur [14], which serves as a defense against predators. When you bite into a radish and chew on it, this mustard oil isothiocyanate is enzymatically released from mustard oil glycoside. Quite a hot thing!

References

1. T. Carow, *Fleischfressende Pflanzen*, 1st edn., Franckh Kosmos Verlag, Stuttgart, **2005**, pp. 6–9.
2. M. Keil and B. P. Kremer (Eds.), *Wenn Monster munter werden*, 1st edn., Wiley-VCH, Weinheim, **2004**, pp. 77–85.
3. T. Carow, *Fleischfressende Pflanzen*, 1st edn., Franckh Kosmos Verlag, Stuttgart, **2005**, pp. 8–9 and 30–33.
4. J. M. Berg, J. L. Tymoczko, G. J. Gatto jr. and L. Stryer, *Stryer Biochemie,* 8th edn., Springer Spektrum Verlag, Heidelberg, **2018**, pp. 53–55.
5. T. Carow, *Fleischfressende Pflanzen*, 1st edn., Franckh Kosmos Verlag, Stuttgart, **2005**, pp. 8–9 and 28–29.
6. Y. Forterre, J. M. Skotheim, J. Dumais and L. Mahadevan, *How the Venus flytrap snaps*, Nature, 433, **2005**, pp. 421–425.
7. S. Feil, *Ether unterbricht Signalweiterleitung bei der Fliegenfalle*, Chem. Unserer Zeit, 56, **2022**, pp. 282–283.

8. J. M. Berg, J. L. Tymoczko, G. J. Gatto jr. and L. Stryer, *Stryer Biochemie,* 8th edn., Springer Spektrum Verlag, Heidelberg, **2018**, pp. 256–297.

9. M. Keil and B. P. Kremer (Eds.), *Wenn Monster munter werden*, 1st edn., Wiley-VCH Verlag, Weinheim, **2004**, pp. 167–182.

10. M. Keil and B. P. Kremer (Eds.), *Wenn Monster munter werden*, 1st edn., Wiley-VCH Verlag, Weinheim, **2004**, pp. 170–172.

11. W. Barthlott and C. Neinhuis, *Purity of the sacred lotus or escape from contamination in biological surfaces*, Planta, 202, **1997**, pp. 1–7.

12. V. Wirth and U. Kirschbaum, Flechten einfach bestimmen, 2., aktualisierte Aufl., Quelle & Meyer Verlag, Wiebelsheim, **2017**, pp. 244.

13. G. Schwedt, *Chemie für alle Jahreszeiten*, 1st edn., Wiley-VCH Verlag, Weinheim, **2007**, pp. 50 and 197.

14. E. Breitmaier and G. Jung, *Organische Chemie*, 7., vollständig überarbeitete und erweiterte Aufl., Georg Thieme Verlag, Stuttgart, **2012**, p. 433.

2

In the Garden, on the Meadow, Terrace or Balcony

Abstract In this chapter, you will learn new things about the green pigment chlorophyll, which is actually much bluer than it appears in leaves. But the yellow and red leaf pigments, which give us the golden autumn, also play an important role, for example as "sunscreen". I try to explain the big topic of photosynthesis in a clear way and with a simple overview graphic. However, you can also conduct exciting experiments with a self-made and long-lasting green chlorophyll solution: it glows blood red when illuminated with light. It no longer glows red when water is added. And with lighter fluid, you can make the yellow autumn pigments visible. The red fluorescence of the green chlorophyll is vividly explained in a swimming pool diving tower graphic. Another topic is climate change and the three most important greenhouse gases. With the climate bottle experiment on the balcony or terrace, you can conduct your own measurements. Finally, it gets fun and colorful with simple experiments: hot grass (no, not

© The Author(s), under exclusive license to Springer-Verlag
Gmbh, DE, part of Springer Nature 2024
A. Korn-Müller, *Scientific Secrets of Nature*,
https://doi.org/10.1007/978-3-662-69575-3_2

cannabis), a stark cress heart, and a natural "fiber optic lamp".

2.1   The Green Leaf Pigment—A Colorful Wonder Bag

The green leaf pigment chlorophyll is easy to isolate from grass and combines chemistry with biology. White light from an LED lamp produces a spectacular red fluorescence in an alcoholic chlorophyll solution, water extinguishes it again. Other green dyes, such as green ink or green food coloring, do not fluoresce. These simple experiments lead to the photochemical principle of photosynthesis.

2.1.1   The Colors of Chlorophyll

Of the six known chlorophyll variants *a–f* so far, chlorophyll *a* and *b* are mainly found in the green plant pigment [1]. The term chlorophyll comes from the Greek and means green *(chloros)* and leaf *(phyllon)*.

Due to its special molecular structure, chlorophyll is an efficient photoreceptor ("light receiver") and is one of the most effective and strongest organic light collecting compounds [1]. Chlorophyll *a* and *b* do not absorb light in the green range, but reflect it, so we perceive their color in our eyes as green. Chlorophyll *a* has absorption maxima at 430 nm (blue) and 662 nm (red), chlorophyll *b* at 453 nm (blue) and 642 nm (red) (1 nm $= 10^{-9}$ m $= 1$ millionth of a millimeter). The resulting leaf color therefore appears green-blue in chlorophyll *a* (Fig. 2.1a) and yellow-green in chlorophyll *b* (Fig. 2.1b).

In a study from 2015, isolated chlorophyll *a* and *b* were measured independently of the solvent with laser

a
Chlorophyll a

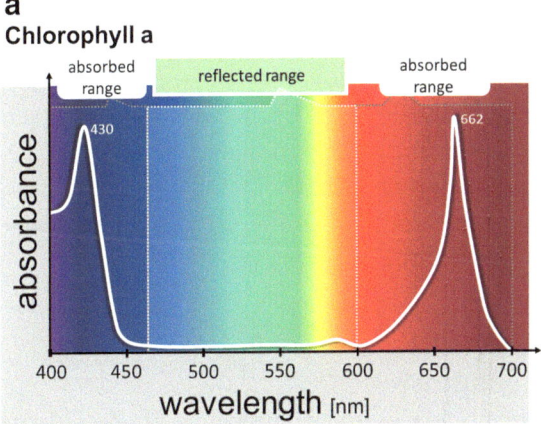

b
Chlorophyll b

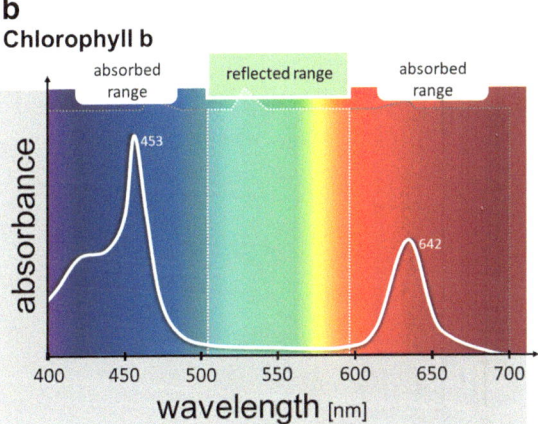

Fig. 2.1 **a** Absorption spectrum of chlorophyll *a*. **b** Absorption spectrum of chlorophyll *b*. Graphic: Melvin Müller

excitation in vacuum [2]. The result: the absorption maxima are shifted by 15–60 nm into the shorter-wave spectrum—in chlorophyll *a* they are at 372 nm and 642 nm respectively, and in chlorophyll *b* at 392 nm and 626 nm. Isolated chlorophyll—without the cellular environment of the photosynthesis proteins and outside the shell (membrane) of the chloroplasts—therefore appears more blue-tinged. We would actually perceive the leaves and grasses as bluer than usual.

While leaves or other green solids, such as green paper, involve reflection of green light, the green chlorophyll solution in glass or plastic containers is referred to as transmission (permeability) of the green light components, as Fig. 2.2 illustrates.

2.1.2 The Large Color Palette

The membrane of the chloroplasts, those cell organelles where photosynthesis takes place, is not only "stuffed" with chlorophyll, but also contains a whole army of other light-sensitive pigments [4]. These include the yellow to red colored carotenoids, such as β-carotene, which gives carrots and pumpkins their orange color, or the bright red

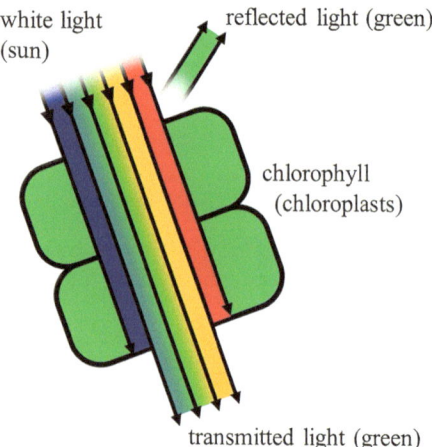

Fig. 2.2 Light path through a green leaf (reflection) or chlorophyll (transmission). Red and blue light is absorbed. (After [3]). Graphic: Melvin Müller

lycopene. Carotenoids are elongated, long-chain, hydrophobic, fat-soluble hydrocarbons that can anchor well and firmly into the membrane. They help with light collection and absorb light of wavelengths between 400 and 500 nm, exactly in the gap that is not covered by chlorophylls *a* and *b* (Fig. 2.3). In addition, the radiation intensity of sunlight is highest at about 500 nm [5]. Xanthophylls are oxygen-enriched (oxidized) carotenoids and are also contained in leaves as color-giving pigments [6, 7]. Lutein is the most common orange-yellow leaf dye from this substance class.

With the help of carotenoids and xanthophylls, the plant loses less light and can capture enough light even under cloudy skies. In addition, carotenoids act as a kind of "sunscreen". If the sunlight is too intense, harmful oxygen can be formed in the cells of the leaves through

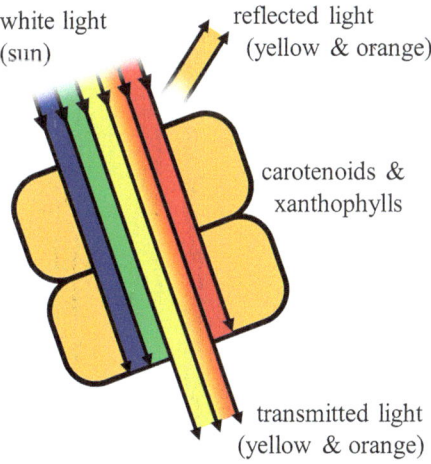

Fig. 2.3 Light path through a yellow leaf (reflection) or carotenoids & xanthophylls (transmission). Carotenoids and xanthophylls absorb green-blue light and therefore appear as yellow-orange-red. Graphic: Melvin Müller

a photochemical reaction from normal atmospheric oxygen. This so-called singlet oxygen is very reactive and can paralyze the entire photosynthesis apparatus through oxidation [4]. You can imagine this as if the leaf were "rusting". Oxidation simply means: oxygen uptake. As with the burning of fuels or the rusting of iron. As a result of oxidation, the leaves would die. Carotenoids prevent this in two ways: First, they absorb much of the light, so that singlet oxygen is not formed in the first place. Second, they convert the energy-rich singlet oxygen into normal oxygen by taking up its energy. The carotenoids release the absorbed energy as heat. It is said that they act as quenchers. But sunscreen sounds much better and more illustrative. In the golden autumn, the chlorophylls are gradually broken down and the carotenoids and xanthophylls become more and more visible, as Fig. 2.4a shows. Also the purple-colored anthocyanins (Fig. 2.4b), which also give many fruits their reddish color, such as red currants, blackberries, blueberries and cranberries but also radishes, eggplants and red cabbage. All three pigments—carotenoids, xanthophylls and anthocyanins—are responsible for the golden yellow to red color splendor and also act as a

Fig. 2.4 **a** Ginkgo tree in October. **b** Amber tree in October

protective factor against too much sunlight. By the way, plants without carotenoids die off quickly [8].

2.1.3 Photosynthesis

In simple terms, you can imagine photosynthesis as a 4 × 100 m relay race. Over several stations (molecules), a baton (electron) is transported as quickly as possible from here to there. The light gives the starting signal and at the finish, the electron reaches the carbon dioxide, which is processed into sugar (glucose).

For those interested, we will now go into a bit more detail about the biochemical nitty-gritty. If this is too much theory for you, you can simply skip this section and start with the chlorophyll experiments from Sect. 2.1.4.

In photosynthesis, two interconnected so-called photosystems (PS) play the main role. Ultimately, it is about the conversion of light energy into chemical energy in the form of molecules. The two photosystems I and II consist of 50–100 molecules, each with a reaction center that is surrounded, indeed completely encircled, by numerous light-harvesting complexes (LHCs). The LHCs lie close together and are studded with chlorophyll and carotenoid molecules that are firmly anchored to membrane proteins. The rigid fixation of all involved pigments is extremely important so that the transmission paths are short and effective. When light hits the photosystems, a whole cascade of reactions is triggered [1, 9, 10].

Everything begins, for historical reasons of discovery, with Photosystem II (PS II). When light of wavelength ≤ 680 nm (red light) hits the green leaf, it is captured (absorbed) by the light-collecting substances of the light-harvesting complexes (LHC) and its energy is transferred losslessly at high speed from molecule to molecule to the center to the reaction center. The light energy is virtually concentrated. So you can imagine the light-harvesting complexes as molecular magnifying glasses that bundle the light and then focus it into the reaction center.

At the center sits a "special pair" of two chlorophyll-*a* molecules, which release an electron when light is absorbed. Electrons are negatively charged elementary particles. The negative electron is absorbed in several stages by electron-loving molecules, "packaged" and gradually sent on its journey over seven stages to Photosystem I. Since PS II is now missing an electron, it gets a positive charge and becomes P680+. This is therefore referred to as light-induced charge separation. These processes happen with unimaginable speed in the billionth to trillionth of a second range! Unimaginable! The P680+ is so energy-rich that it quickly retrieves its lost electron by stealing the electrons from a water molecule. In the process, water is split and converted into oxygen gas. In total, four electrons are removed from two water molecules, the P680+ is reset to the original P680 and the oxygen is released into the air, which we then breathe. Four protons, i.e., positively charged hydrogen atoms (H+), are also produced. They are important later. The reaction center of Photosystem II is named P680 because it works with light of wavelength ≤ 680 nm, and P stands for the "special *P*air".

The wild electron ride continues in Photosystem I and its reaction center P700. This is also surrounded by numerous light-harvesting complexes in a ring shape. Light of wavelength ≤ 700 nm (red light) is collected, its energy is transferred, concentrated, and transported to the reaction center. Here too, a "special pair" of two chlorophyll-*a* molecules sits, from which the light knocks out an electron, thus making the P700 positive (P700+). This electron is passed on through five stages by various molecules and finally ends up at NADP+. NADP+ takes up the electron and thus becomes one of the most important biochemical substances in nature, NADPH (Nicotinamide Adenine Dinucleotide Phosphate), which plays a central role in the production of complex, organic substances, such as proteins, carbohydrates, fats, nucleic acids, and hormones. The P700$^+$ gets its lost electron back from Photosystem II, and so the entire electron flow ends like a domino cascade. The released protons (H$^+$) are used for the production of ATP (Adenosine Triphosphate), the most significant biochemical energy carrier in nature [1, 9, 10].

And what is the bottom line? Photosynthesis converts light energy into chemical energy. This results in NADPH, ATP, and oxygen. All of this happens during light exposure and is therefore called the light reaction. The two chemical energy carriers NADPH and ATP then flow into the so-called dark reaction (Calvin cycle), in which carbon dioxide from the air is converted into glucose in several steps with the help of NADPH and ATP [11]. The plant does not need light for these processes, hence the name dark reaction. From glucose, the plant can finally form starch and other carbohydrates. Figure 2.5 summarizes the results graphically. Conclusion: Photosynthesis is a prime example of the interplay of physics, chemistry, and biology.

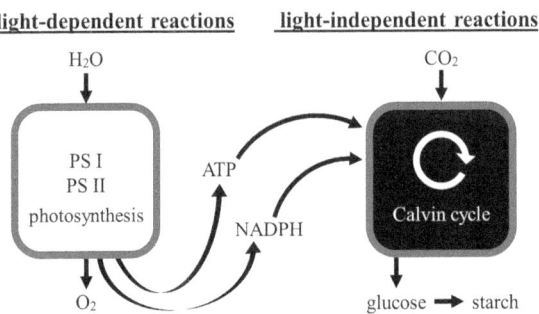

Fig. 2.5 Schematic of the light (light dependent) and dark (light independent) reaction in the chloroplasts. Graphic: Melvin Müller

2.1.4 Experiment: Making a Chlorophyll Solution

Chlorophyll is easy to extract, and the alcoholic extraction is also suitable for students from the 5th grade or from about 9 years old with the help of the teacher or parents [12, 13].

You will need

- Meadow, lawn
- Ethanol (96 Vol.-%) or denatured alcohol
- Scissors
- Jam jar with lid, screw-top jar approx. 500–750 mL volume
- Empty PET bottle (500 mL volume)

How it works
About a handful of fresh grass (approx. 10 g) is cut into small snippets with scissors and put into a well-sealing screw-top jar with 500–750 mL volume (Fig. 2.6). Now add 200 mL of ethanol (96 Vol.-%) or denatured alcohol,

Fig. 2.6 Extraction and filtration of chopped grass using ethanol. **a** Small cut grass snippets in a screw-top jar. **b** Filtration of the ethanol-grass mixture. **c** Chlorophyll solution as filtrate in a small PET bottle

close the jar and shake it a few times. After about 20–30 min, filter the green solution through a coffee filter into a 500 mL PET bottle (Fig. 2.6). The chlorophyll solution can be stored in a closed container, protected from light, for months. It gradually fades in light.

2.1.5 Experiment: Red Glow in the Green Grass—Tracking Photosynthesis with a Flashlight

A first experiment on the topic of chlorophyll/leaf green could start by having the children or friends as a "research task" illuminate a meadow or a lawn area in the dark with a flashlight to test whether a red glow occurs. The main experiment can then start the series of experiments with the chlorophyll solution.

You will need

- Alcoholic chlorophyll solution (Sect. 2.1.4)
- Narrow glass (tall drinking glass, test tube)
- LED flashlight or smartphone light
- If available: flashlight with blue filter or UV lamp
- Darkness

How it works

If you pour the ethanolic chlorophyll solution into a narrow glass and illuminate it perpendicularly from below, no fluorescence can be seen, the chlorophyll solution appears green. However, if the solution is illuminated from the side, a red fluorescence can be observed (Fig. 2.7a and b). The red fluorescence is clearly visible with an LED flashlight or the light of a smartphone. Using a blue filter enhances the effect even more, as the entire light consists of short-wave and thus more energy-rich light. In addition, the perception of fluorescence is improved. The red fluorescence appears particularly spectacular in a round or Erlenmeyer flask with a wide liquid surface, as shown in Fig. 2.7c and d.

What's the science behind it?

The fluorescence light behaves isotropically, that is, it radiates evenly in all directions. If the excitation light falls on the solution from a horizontal direction, almost no light from the excitation lamp enters the observer's eye. The fluorescence light, which spreads in all directions, can therefore be seen on the surface of the solution (Fig. 2.7). However, if the glass is illuminated from below, observers face the excitation light. The red fluorescence light still reaches the eye, but it is outshone by the bright excitation light, as it is more intense [13].

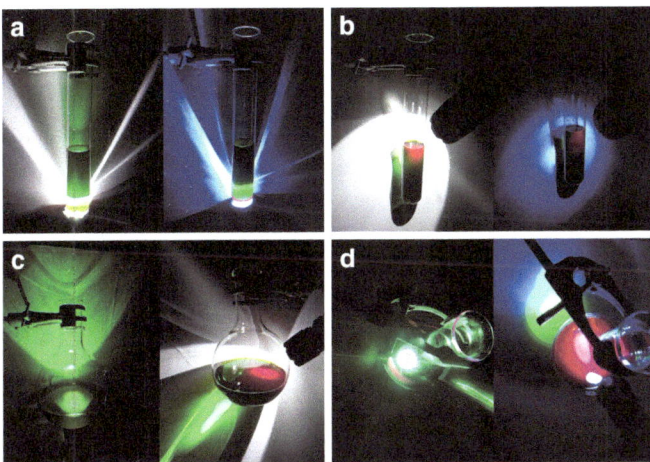

Fig. 2.7   a Chlorophyll solution with white light (left) and blue light (right) illuminated perpendicularly from below. The solution appears green. **b** The same solutions with white light (left) and blue light (right) illuminated from the side. The solution fluoresces red. **c** Chlorophyll solution in a round flask with white light illuminated perpendicularly from below (left) and from the side (right). **d** The same solution with blue light illuminated perpendicularly from below (left) and from the side (right)

Through absorption of light, electrons in the photosynthesis apparatus move from the ground state to the excited state. Normally, the photosynthesis substances anchored in the chloroplast membrane absorb these electrons and pass them on in a cascade-like manner. In this process, about 80% of the absorbed light energy is used photochemically, 19.5% is released as heat to the environment, and 0.5% is lost as fluorescence light [14]. Since the chlorophyll solution does not contain chloroplasts or these are only present in a destroyed (denatured by ethanol) form, this process no longer works. The electrons lose up to 97.5% of the energy as heat within picoseconds without radiation and then return from the excited state to the ground state.

In doing so, they release the remaining energy as longer-wave light in the form of a red fluorescence with an emission maximum of 685 nm to the environment [14, 15]. A green leaf, a meadow, or green vegetables illuminated with white or blue light also fluoresce, but the fluorescence only accounts for 0.5% of the incident light and is not visible or perceptible to our eye. The red chlorophyll fluorescence of (crop) plants is also used to determine the "health status" of the respective plant. This phenomenon, known for decades, allows indirect measurement of stress factors such as lack of nutrients, heat, water shortage, or pest infestation [14]. The worse a plant is doing, the less light energy it can convert into chemical energy, and the brighter the fluorescence shines. Today, the fluorescence spectra are recorded from masts or airplanes. In the future, a fluorescence satellite of the ESA is supposed to provide high-resolution fluorescence maps, for example, to research crop failures or global vegetation in climate change. The graphic in Fig. 2.8 illustrates the red fluorescence of green chlorophyll once again.

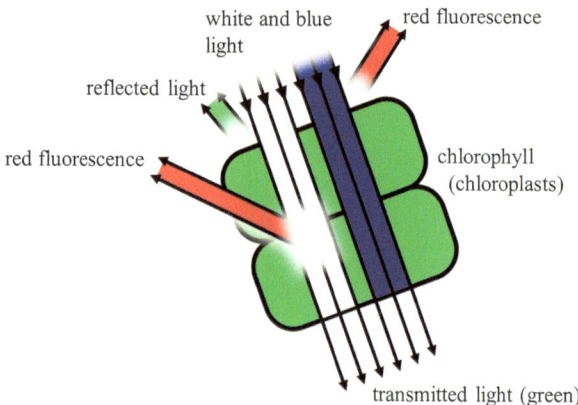

Fig. 2.8 Red fluorescence of green chlorophyll when irradiated with white or blue light. Graphic: Melvin Müller

Finally, the question can be answered why a meadow does not appear to glow red to our eye at night when illuminated with white or blue light. The cells of the grass blades—like in all other green leaves—still contain the chloroplasts, the intact proteins of the light-harvesting complexes, and the two photosystems. These molecules largely feed the electrons into the photosynthesis cascade without loss and "utilize" them.

2.1.6   Experiment: Making Fluorescence "Disappear"—The Quenching Effect

You will need

• Alcoholic chlorophyll solution (Sect. 2.1.4)
• Water
• Dish soap
• Plastic syringe (10–20 mL, pharmacy, hardware store)
• Narrow glass (tall drinking glass, test tube)
• LED flashlight or smartphone light
• Darkness

How it works
Add about 10 mL of chlorophyll solution to a narrow glass using a plastic syringe and illuminate the solution from the side in the dark: The red fluorescence is clearly visible. Now add water milliliter by milliliter with the syringe. The glass is continuously illuminated during this process. From a certain amount of water (approx. 7 mL), the red fluorescence disappears and the solution only glows greenish (Fig. 2.9).

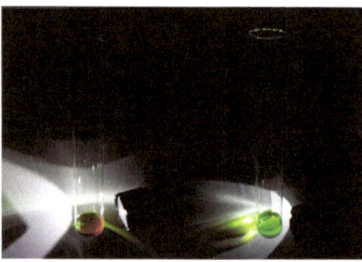

Fig. 2.9 Illumination of a chlorophyll solution with white light from the side. **a** Initially, the red fluorescence appears. **b** After adding water, the red fluorescence disappears

The disappeared fluorescence can be brought back: Add a strong "shot" of dish soap to the glass. Swirl. Illuminate. If nothing is visible yet, simply add more dish soap. With side illumination, the red fluorescence reappears—but quite weak and not as colorful as at the beginning.

What's the science behind it?

The quenching of fluorescence, the so-called quenching effect, is based on the aggregation of chlorophyll molecules in a hydrophilic environment (English *to quench* = extinguish). This means that the rather hydrophobic (water-repellent) chlorophyll molecules "huddle" together in water, aggregate, "clump" together, so to speak, and pass on the light energy among themselves. There is nothing left for fluorescence.

Dish soaps contain surfactants, which consist of a hydrophilic (water-loving) head and a hydrophobic (water-repellent) tail. You can imagine their structure like a match: the head loves water, the tail avoids water. Such surfactants form microscopic small spheres, so-called micelles, in water. The heads are located on the surface, while the tails form the sphere closely packed together. When cleaning, the hydrophobic tails pick up grease and dirt and encapsulate them. In our experiment, instead of grease or dirt, the chlorophyll molecules are incorporated into the micelle membrane, as chlorophyll is a hydrophobic substance. Thus, individual chlorophyll molecules can be "fished out", "captured" and isolated from the water-ethanol mixture. And they are again suitable for fluorescence. Presumably, not all chlorophylls are incorporated into the micelles. This would explain why the red fluorescence does not reappear in full splendor.

2.1.7 Experiment: Do Other Green Dyes also Glow Red?

To investigate whether other green dyes also fluoresce red when illuminated, solutions are prepared from, for example, green ink or green food coloring [13].

You will need

- Green ink (fountain pen ink dark green 4001, manufacturer: Pelikan, 6 cartridges = 1.49 €) or
- Green food coloring (E131 = Patent Blue V, manufacturer: Rosenheimer Gourmet Manufaktur; 40 mL = 2.99 €) or
- Green food coloring (E141 = Copper Chlorophyllin, manufacturer: Wusitta; 20 mL = 0.99 €)
- Ethanol or denatured alcohol
- Glass (for mixing)
- Narrow glass (tall drinking glass, test tube)
- LED flashlight or smartphone light
- Small measuring cup or syringe (50 mL, pharmacy, hardware store)

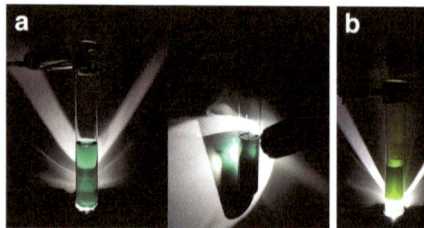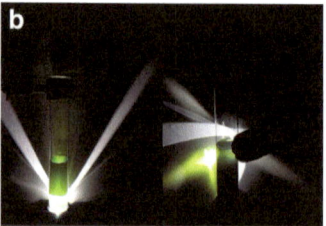

Fig. 2.10 **a** Irradiation of a green ink solution with white light from below and from the side. **b** Irradiation of a solution with green food coloring with white light from below and from the side. No red fluorescence can be seen in any of these green solutions

How it works

Add 2–10 drops of a green dye to 50 mL of ethanol or denatured alcohol. From the green color solutions, add 20–30 mL each to a narrow glass and illuminate from below and the side in the dark. Figure 2.10 shows the behavior of green ink or green food coloring when exposed to light. None of the green solutions show a red fluorescence.

What's the science behind it?

The green dye molecules used are photochemically inactive and cannot be excited like chlorophyll. This makes chlorophyll so unique and indispensable for photosynthesis.

2.1.8 Experiment: Making the Yellow Leaf Pigments Visible: Green on Top, Yellow Below

Our alcoholic chlorophyll solution (Sect. 2.1.4) is a wild mix of numerous leaf pigments, namely: the green chlorophyll molecules and many other yellow carotenoid and xanthophyll pigments. With a simple experiment, the green pigments can be separated from the yellow ones and made visible [16].

You will need

- Alcoholic chlorophyll solution (Sect. 2.1.4)
- Water
- Lighter fluid (supermarket)
- Small measuring cup
- Narrow glass (tall screw-top jar, test tube with stopper)

How it works

Pour 20 mL of alcoholic chlorophyll solution into a narrow screw-top jar or test tube and add another 10 mL of water. This results in a clear, green solution. Now add 20 mL of lighter fluid to the green solution, close the jar with the lid or stopper, and shake the contents vigorously. Since water/ethanol and gasoline do not mix, a cloudy emulsion forms, which separates into two layers after a few minutes. The upper layer (also referred to as phase) contains the lighter gasoline with the dissolved green chlorophyll dyes. The lower layer is heavier and consists of aqueous ethanol with dissolved yellow carotenoid and xanthophyll dyes (Fig. 2.11).

What's the science behind it?

Gasoline is absolutely nonpolar and thus highly water-repellent. It does not mix with the polar water, separates from it, and "floats" on top of the water due to its lower density. Density of water: 1.0 g/cm^3, density of gasoline:

Fig. 2.11 Separation of green and yellow leaf dyes of an alcoholic chlorophyll solution, which was mixed with lighter fluid. (After demixing, the green chlorophyll pigments are located in the upper, lighter gasoline layer, while the yellow carotenoid and xanthophyll dyes are dissolved in the lower, heavier water-alcohol phase)

0.68 g/cm^3. Water is therefore about 1.5 times heavier than gasoline. The chlorophyll molecules are also very nonpolar and water-repellent and therefore migrate from the alcoholic solution to the gasoline into the gasoline phase. They simply feel more comfortable there. The yellow leaf dyes such as the carotenoids and xanthophylls (lutein) are rather polar and water-soluble. They remain in the ethanol-water phase [16].

Background

The Fluorescence

Fluorescence refers to the emission of light that only lasts as long as the fluorescing molecules are excited by light absorption. With the help of light, electrons of a fluorescence-capable molecule absorb energy and are transported in an extremely short time of 10^{-13} s from the ground state (S_0) to an excited singlet state (S_2). Within 10^{-11} s, radiationless vibrational relaxation and thermal equilibration occur through molecular collisions, resulting in a release of energy in the form of heat to the surroundings. Through this *internal conversion*, the excited molecule finally falls to the lowest energy excited singlet state S_1. From there, the remaining part of the excitation energy is emitted in the form of visible light within 10^{-8} s during the transition from S_1 to the ground state S_0 and is referred to as fluorescence [17, 18]. Due to the energy loss, fluorescence light is shifted towards longer-wavelength light compared to the shorter-wavelength excitation light (usually UV light). Fluorescence only occurs during simultaneous illumination and is therefore a simultaneous light emission when irradiated with light. Common fluorescence phenomena in everyday life include, for example, the highlighter pens glowing under UV light, banknotes, esculin from the branches of the horse chestnut, certain fungi, yellow lichens on trees, or overripe banana peels.

So, that was the chemically and physically correct explanation of fluorescence. With Fig. 2.12, however, I would like to try to illustrate the energy scheme of fluorescence in a vivid, comic-like way.

And it goes like this: A brave, green electron climbs with a lot of energy up to the 10 m board of the diving tower. But there is too much crowding and shoving up there, and it doesn't quite dare to jump down, so the electron climbs back down. This uses up some energy and it now reaches the 3 m board with less energy. This is a

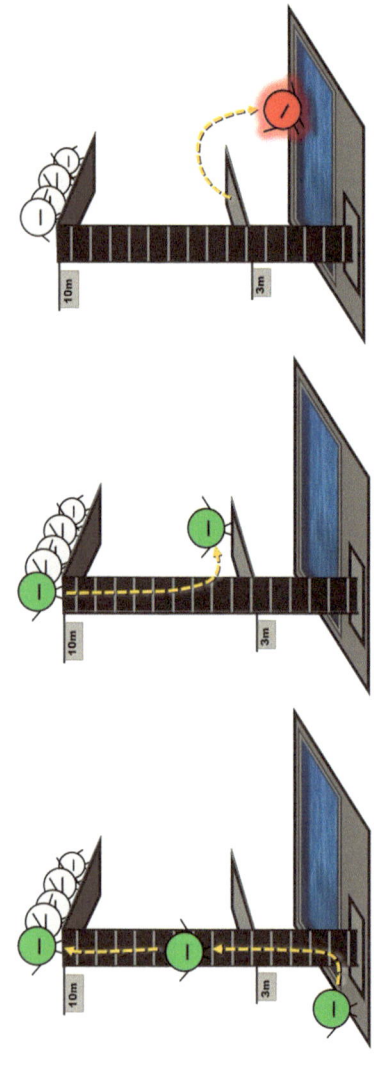

Fig. 2.12 Illustration of fluorescence, vividly represented as a diving board. Graphic: Melvin Müller

comfortable height and the electron jumps from there into the pool. With a shout of joy and a red face, it releases its remaining energy. So we see the red glow of the head or the "light splashes" of an electron that has jumped into the water.

Learn more about other spectacular fluorescence phenomena in nature in Chaps. 5 and 6.

2.2 The Greenhouse Effect on Earth

What some people may not know: The greenhouse effect is initially very positive and is mainly caused by the invisible water vapor (H_2O) present in the air, accounting for about 36–70%. Approximately 9–26% is due to carbon dioxide (CO_2), and about 3–9% each is caused by methane (CH_4), nitrous oxide (N_2O), and natural ozone (O_3) [19]. About 70% of the light radiation hitting the earth's surface from the sun passes through the atmosphere. On the surface, most of it is converted into heat radiation (infrared light), which is absorbed by the gas particles in the air. In particular, the small molecules listed above, with their asymmetric charge distribution, are perfectly capable of absorbing the infrared radiation and reflecting it back towards the earth. This leads to warming (greenhouse effect). The average temperature on earth is a comfortable +14 °C. Without the greenhouse effect caused by water vapor and CO_2 that has been present for millions of years, the earth's temperature would be a chilling −18 °C [19]. So far so good. But now the man-made greenhouse effect is added. Here, three gases play a major role, carbon dioxide, methane, and nitrous oxide, because only three- or more-atomic molecules can cause a greenhouse effect. Nitrogen (N_2, 78% in the air) and oxygen (O_2, 21% in the air) are diatomic gases and contribute nothing to the greenhouse effect, as they are only capable of symmetric stretching vibrations and are therefore not infrared active. The main cause of the man-made temperature increase is carbon dioxide (CO_2), which can be excited by infrared radiation to deformation vibrations. These asymmetric vibrations release heat energy. The same happens with water (H_2O), methane (CH_4), and nitrous oxide (N_2O). Most greenhouse gases come from fossil energy, livestock farming, deforestation, soil cultivation (fertilization), and biomass combustion [20]. In Table 2.1, the most important facts are listed again.

Table 2.1 Greenhouse gases in the atmosphere (selection), status 2022. (After Lit. [20])

gas	molecule	concentration	residence time in the atmosphere (approx.)	percentage of the natural greenhouse effect (approx.)	percentage of the anthropogenic greenhouse effect (approx.)
air water vapour	N_2, O_2, H_2O	/	10 days	60 %	/
carbon dioxide	CO_2	420 ppm	120 years, 85-60 % degradation after 1,000 years	26 %	58 %
methane	CH_4	2,000 ppb	9 years	2 %	22 % (25x more effective than CO_2)
laughing gas	N_2O	332 ppb	131 years	4 %	6 % (300x more effective than CO_2)

Background

Greenhouse gases

Rank 1: Carbon dioxide (CO_2).

Worldwide, human-made emission of CO_2 *per day*: approx. 100 million tons (as of 2022)! Although CO_2 currently only makes up 0.042% ($= 420$ ppm) of the air, it has a massive, global impact on global warming and thus on the global climate (2–3 ppm annual increase, 0.2 °C warming every ten years). 0.042% CO_2 corresponds to a roughly 20 m thick gas layer of pure CO_2 around the globe. Oxygen with 21%: 10 km thickness, nitrogen with 78%: 39 km thickness (at 50 km height of the atmosphere and an earth surface of 2.55×10^{10} km²). The small amount of carbon dioxide should not deceive us that the global climate reacts extremely sensitively and even minor changes have dramatic effects on it, which has been in equilibrium for millions of years. In 1800, the value was still at 280 ppm, in 1965 at 320 ppm, in 2005 at 380 ppm, and in 2021 at 417 ppm.

Rank 2: Methane (CH_4).

It rose from 1700 ppb in 1988 to 2000 ppb in 2021. It is about 25 times more effective as a greenhouse gas than CO_2.

Rank 3: Nitrous oxide (dinitrogen oxide, N_2O).

It is about 300 times more effective than CO_2. It rose from 300 ppb in 1978 to 332 ppb in 2020. About 0.8 ppb increase per year.

ppm (parts per million) = 1 millionth part;

e.g., 1 mL ($= 1$ cm³) in 1 m³ ($= 1000$ L $= 1$ million cm³);

corresponds to 1/10,000th percent.

400 ppm CO_2 means: 0.04% CO_2 or 400 CO_2 molecules per million molecules of dry air.

ppb (parts per billion) = 1 billionth part;

e.g., 1 mL ($= 1$cm³) in 1000 m³ ($= 1$ million L $= 1$ billion cm³);

corresponds to 1/10,000,000th percent.

400 ppb CH_4 means: 0.00004% CH_4 or 400 CH_4 molecules per billion molecules of dry air.

The IPCC (Intergovernmental Panel on Climate Change, founded in 1988) is tasked with compiling, evaluating the global state of research and scientific findings, and assessing the resulting risks and impacts of global warming. The IPCC has calculated four possible future scenarios for anthropogenic, i.e., human-made climate change, up to the year 2100 [21]. The basis is tiered emission values of carbon dioxide and other greenhouse gases, which are converted and included as CO_2 equivalents. These four scenarios are referred to as RCP (Representative Concentration Pathway) and are as follows [22].

RCP 2.6 ("We stop global CO_2 emissions immediately"): At 490 ppm (= 0.049%) CO_2 as the best, most optimistic scenario, the Earth would warm by a maximum of 1.7 °C and raise the sea level "only" by up to 40 cm.

RCP 4.5: At 650 ppm (= 0.065%) CO_2 as a moderately good scenario, the temperature would rise by a maximum of 2.6 °C and raise the sea level by at best 47 cm.

RCP 6.0: 850 ppm (= 0.085%) CO_2 as a bad scenario would result in a global warming of up to 3.1 °C and raise the sea level to up to 63 cm.

RCP 8.5 ("We continue to live as before"): With 1370 ppm (= 0.137%) CO_2 as the worst case, the Earth would warm by 4.8 °C and the sea level would rise by up to 92 cm [22].

The Mauna Loa Observatory with its Global Monitoring Laboratory in Hawaii has been measuring the CO_2 content in the atmosphere daily since 1980 and can be followed in real-time on the Internet at any time: https://gml.noaa.gov. On this highly recommended website, you will find numerous data, measurement results, evaluations, and great animations on all greenhouse gases and climate changes on our Earth.

2.2.1   Experiment: Greenhouse Effect in a Bottle

With a fairly simple experiment, you can recreate and visualize the greenhouse effect on a small scale.

You will need

- 3 PET bottles (500 mL volume, cleaned and dried) with lids
- 3 Thermometers that fit into the bottles (hardware store)
- Some black paint (acrylic, tinting paint, hardware store) or a piece of black fabric
- Carbon dioxide capsules (soda capsules)
- Nitrous oxide/cream gas capsules (cream spray gas from a cream dispenser)

How it works

To ensure sufficient conversion of light into heat radiation, the bottom of each bottle is painted black from the inside. Simply put a blob of paint inside and spread it with a long brush. Alternatively, you can also put a piece of black fabric into the bottles. The bottom of the bottle should be covered.

The first bottle is filled only with normal air, i.e., you take fresh air from outside by waving the empty bottle back and forth. Then you put the thermometer in the bottle and screw on the lid. Label the bottle or the lid with "Air".

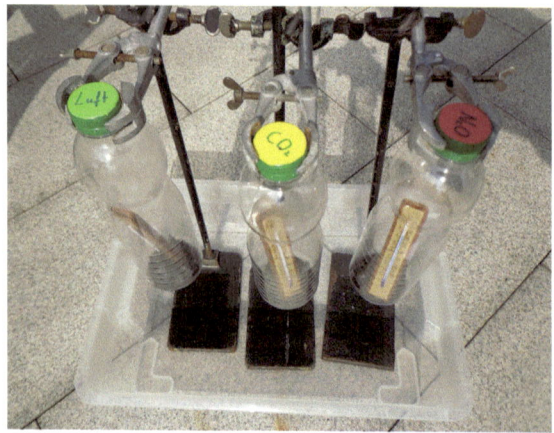

Fig. 2.13 Three gas-filled "greenhouse bottles" equipped with a thermometer. Green painted lid: Air, yellow lid: CO_2, red lid: N_2O

The second and third bottles are filled with nitrous oxide (dinitrogen oxide, N_2O) or CO_2, respectively. Nitrous oxide is the "cream gas" used to foam liquid cream in a cream dispenser. To do this, insert a nitrous oxide or CO_2 cartridge into the cream dispenser and spray the pure gas into the respective bottle. To completely displace all the air from the bottle, repeat the filling with the gas once more. Then the thermometers are inserted, the bottles are closed and labeled. The three filled bottles are now positioned on the south or sunny side on the terrace or balcony. Figure 2.13 shows my three gas bottles on the terrace. By the way, in the air bottle, water vapor forms in direct sunlight, which condenses on the edge of the bottle and thus makes reading the thermometer a bit difficult.

What's the science behind it?
Especially in high summer with strong sunlight, you get a clear result. In cloudy or cold weather without sunlight, the temperatures in all bottles are roughly the same. It

is a very simple but impressive experiment. In the CO_2 atmosphere, the temperature value is about 1–2 °C higher compared to the temperature in the air bottle, and even a significant 3–4 °C higher with nitrous oxide. However, one must consider that in the natural atmosphere there is neither 100% CO_2 nor 100% N_2O as in the bottles, but only amounts in the ppm or ppb range. But that is enough to knock the Earth's fragile climate system off balance. The greenhouse effect is much more complex and also depends, among other things, on the reflection of sunlight from the surface. This reflectivity is called albedo. You all know the absorption capacity of heat radiation by different materials: A black painted car gets significantly hotter in the sun than a white car. Snow and ice reflect sunlight much better than brown earth, red desert sand or gray rocks. The oceans have a very small albedo value because water has an incredibly large heat capacity. No substance in the world can absorb heat as well as water. Water is the world champion in cooling because it can absorb so much heat. That's why the fire department likes to extinguish with water. Even with boiling hot water, you can still put out a fire. Burned your finger? Put it under the faucet! Great idea! Go to the world champion of cooling. The downside of all this: The world's oceans are increasingly warming up with fatal consequences for flora and fauna and especially for the polar ice regions. While ice and snow reflect up to 90% of sunlight back into space, water has a reflectivity of only about 5–20%. This means: 80–95% of the radiation energy remains in the water [23]. Over the decades, a self-perpetuating vicious circle of melting ice, water formation and increasing warming has formed, which was declared a tipping point in 2022 [23]. This process is irreversible.

2.3   Some Like It Hot

The ground is teeming with tiny organisms everywhere, bacteria, fungi, and small animals. There are more organisms in a tablespoon of forest soil than there are people on Earth! Hundreds of species frolic there, decomposing dead leaves, needles, branches, etc. into nutrients and humus. Just like in a compost heap. During decomposition by bacteria, energy is also produced in the form of heat [24]. You can easily investigate and measure this with a simple experiment.

2.3.1   Experiment: Heat from Gas? No, Heat from Grass!

If you are the lucky owner of a meadow or a lawn, you can try this experiment the next time you mow the lawn in summer.

You will need

- Meadow and lawnmower
- Bio-waste bag (compostable)
- Thermometer (hardware store)

How it works
Mow the lawn and stuff the cut grass into a bio-waste bag. The grass should be fresh and nicely packed. Now stick a thermometer into the pile of grass, as shown in Fig. 2.14b. The temperature rises to over 30 °C in a short time.

Fig. 2.14   a The outside temperature is 22 °C. **b** After 20–30 min, the inside of the grass clippings in the bag heats up to 35 °C

What's the science behind it?

Like the forest floor, meadows, grasses, and flowers are also teeming with bacteria. As soon as the grass is cut, the bacteria begin to decompose the stems. In doing so, they "burn" the plant glucose and cellulose with atmospheric oxygen to carbon dioxide. Since this reaction is exothermic, energy in the form of heat is released to the environment. Ultimately, it is bacterial enzymes that break down and utilize the plant's constituents. Moisture is important because water assists in the decomposition process.

To prevent freshly mowed hay from rotting, it is turned with a turning device over many days, loosely spread on the meadow, and air-dried. Only dry hay resists decomposition and can later be used as horse or livestock feed. Entire barns have even burned down because the moist grass piles inside heated up so much that their insides dried out and caught fire.

By the way: Freshly mowed grass always smells so nice. The corresponding fragrance is called coumarin and is also found in woodruff.

2.4   Radical Cress

If you need a creative idea for your loved ones—here's my offer: With cress seeds, you can grow any words or figures as green plants [25]. It's uncomplicated and relatively quick for biological standards. You can present a cress heart to your loved one, for example, on Valentine's Day with the words: "I love you crass!" Then turn off the light and turn on the UV lamp—the cress heart glows in the most beautiful red. In addition, cress is very healthy, contains many nutrients, such as iron, calcium, folic acid, and vitamin C. With its sharp-spicy fresh taste, it refines soups, salads, pasta, and fish dishes.

2.4.1   Experiment: Cress Heart—I Love You Crass!

You will need

- Cress seeds
- Kitchen roll paper
- Water sprayer
- Plate

How it works
Place three layers of kitchen roll paper on a large plate and moisten them with water using a flower sprayer. The absorbent paper should be nicely wet. Now the seeds are sprinkled as letters or symbols on the wet absorbent paper. The more precisely you arrange the seeds, the more

Fig. 2.15 Cress seeds on moistened paper towel **a** after two days, **b** after six days

impressive the result will be. I have sown a heart as an example together with my daughter (12). She cut out a heart shape from cardboard and placed it in the middle of the wet kitchen paper. The cress seeds were then sprinkled around the outlines. Afterwards, the cardboard is carefully removed with tweezers and the sowing is done. Now all you have to do is wait and keep the paper nicely moist. Regular spraying with water is absolutely necessary so that the seedlings do not dry out. Fertilizer is not necessary as the plant brings its nutrients in the germ itself. After two days the cress seeds germinate, after four days the young plants appear and from day six you can admire your growth picture. Figure 2.15 shows my daughter's cress heart. Finally, you can cut the cress and enjoy eating it.

What's the science behind it?

Cress is easy to cultivate, it is an undemanding cruciferous plant, does not need fertilizer, germinates and grows very quickly. The typical sharp-spicy taste is due to the substance benzyl isothiocyanate, which is also contained in horseradish and nasturtium. Benzyl isothiocyanate consists of an (aromatic) carbon six-ring and an $N=C=S$ group, which are linked together via a carbon group (CH_2). It has bacterio-static, virostatic, and antimycotic as well as anticancerogenic effects—it practically helps against all possible pathogens. Similar isothiocyanates are also contained as defense and sharp substances in radishes or mustard seeds [26].

2.5 Radical Puzzle

At the end of this chapter, there should be a quiz question. Take a look at Fig. 2.16!

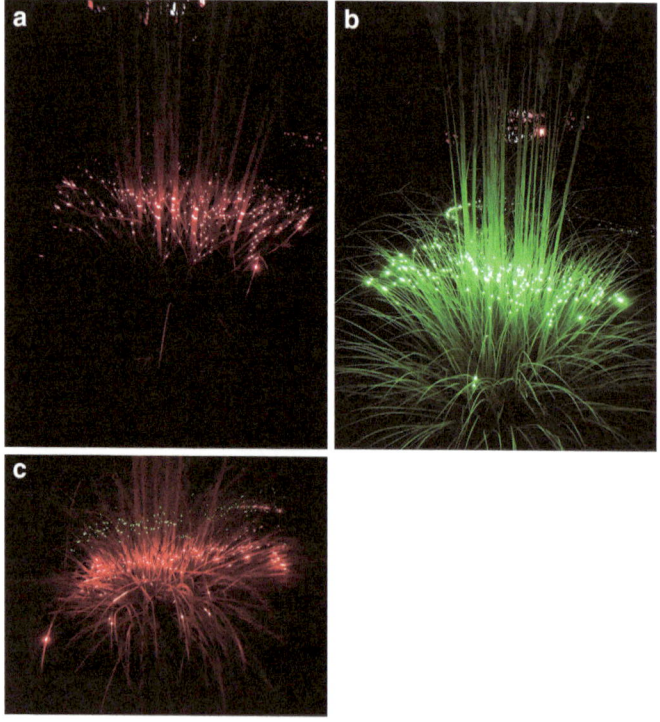

Fig. 2.16 Pampas grass irradiated with laser light from laser pointers: **a** red light, **b** green light, **c** red and green laser beam

What is this?

a) Fiber optic lamp
b) Glowing water fountain in an amusement park
c) Avatar plant

2.5.1 Experiment: Natural Fiber Optic Lamp

You will need

- Green or red laser pointer
- Fine-bushy plant, e.g. pampas grass, grasses

How it works
Wait for darkness and then irradiate the grasses of a pampas grass or any shrubbery with as many stems and stalks as possible with a laser pointer. Hold the laser horizontally at about hip height and "wave" the pointer or the laser beam in quick movements at a vertical angle to the plant between the stalks back and forth, from right to left and vice versa.

What's the science behind it?
Each stalk of the plant reflects the incident laser light and therefore briefly lights up as a green or red dot. With very fast movements of the laser beam, it looks as if numerous stalks are simultaneously lighting up at points. Since each stalk has an arc-shaped shape, the laser beam is reflected several times by the same stalk. The "total work of art" has astonishing similarities with fiber optic lamps.

References

1. a) https://www.bundeswaldinventur.de/dritte-bundeswald-inventur-2012/klimaschuetzer-wald-weiterhin-kohlenstoff-senke/ (last access: Oct 18, 2024). b) https://www.wald.de/waldwissen/wie-viel-kohlendioxid-co2-speichert-der-wald-bzw-ein-baum/ (last access: Oct 18, 2024). c) https://www.sdw.de/ueber-den-wald/waldwissen/wald-in-zahlen/ (last access: Oct 18, 2024).
2. B. F. Milne, Y. Toker, A. Rubio and S. B. Nielsen, *Unraveling the Intrinsic Color of Chlorophyll,* Angew. Chem. Int. Ed. 54, **2015**, pp. 2198–2201.
3. L. Urry, M. Cain, S. Wasserman, P. Minorsky and J. Reece, M., *Campbell Biologie*, 11., aktualisierte Aufl., Pearson Verlag Deutschland, München, **2019**, p. 258.
4. M. T. Madigan, J. M. Martinko, D. A. Stahl and D. P. Clark, *Brock Mikrobiologie kompakt,* 13., aktualisierte Aufl., Pearson Deutschland, München, **2015**, pp. 110–111.
5. P. A. Tipler and G. Mosca, *Physik*, 8., korrigierte und erweiterte Aufl. (Eds.: P. Kersten and J. Wagner), Springer Spektrum Verlag, Berlin, **2019**, p. 654.
6. H. Bannwarth, B. P. Kremer and A. Schulz, *Basiswissen Physik, Chemie und Biochemie*, 4., aktualisierte Aufl., Springer Spektrum Verlag, Berlin, **2019**, 412–414.
7. G. Schwedt, Chemie für alle Jahreszeiten, 1st edn., Wiley-VCH Verlag, Weinheim, **2007**, pp. 55–57 and 202–203.
8. J. M. Berg, J. L. Tymoczko, G. J. Gatto jr. and L. Stryer, *Stryer Biochemie,* 8th edn., Springer Spektrum Verlag, Heidelberg, **2018**, p. 680.
9. L. Urry, M. Cain, S. Wasserman, P. Minorsky and J. Reece, M., *Campbell Biologie*, 11., aktualisierte Aufl., Pearson Verlag Deutschland, München, **2019**, pp. 253–269.
10. M. T. Madigan, J. M. Martinko, D. A. Stahl and D. P. Clark, *Brock Mikrobiologie kompakt,* 13., aktualisierte Aufl., Pearson Deutschland, München, **2015**, pp. 106–118.

11. J. L. Slonczewski and J. W. Foster, *Mikrobiologie*, 2nd edn., Springer Spektrum Verlag, Berlin Heidelberg, **2012**, pp. 624–630.

12. A. Korn-Müller and A. Steffensmeier, *Das verrückte Experimentier-Labor*, 1st edn., Fischer-Sauerländer Verlag, Frankfurt, **2019**, p. 35.

13. A. Korn-Müller, *Warum Gras nicht rot leuchtet*, Nachr. Chem. 70, **2022**, pp. 18–21.

14. H. R. Bolhar-Nordenkampf, S. P. Long and E. G. Lechner, Die Bestimmung der Photosynthesekapazität über die Chlorophyllfluoreszenz als Maß für die Stressbelastung von Bäumen, Phyton (Austria), 29, **1989**, pp. 119–135.

15. G. H. Krause and E. Weis, *Chlorophyll fluorescence and photosynthesis: the basics*, Annu. Rev. Plant Physiol. Plant Mol. Biol., 42, **1991**, pp. 313–349.

16. G. Schwedt, *Chemie für alle Jahreszeiten*, 1st edn., Wiley-VCH Verlag, Weinheim, **2007**, pp. 48–49.

17. E. Breitmaier and G. Jung, *Organische Chemie*, 7., vollständig überarbeitete und erweiterte Aufl., Georg Thieme Verlag, Stuttgart, **2012**, pp. 554–555.

18. D. Weiß and H. Brandl, *Fluoreszenzfarbstoffe in der Natur, Teil 1*, Chem. Unserer Zeit, 47, **2013**, p. 53.

19. https://www.helmholtz-klima.de/faq/was-ist-der-natuerliche-treibhauseffekt (last access: Oct 18, 2024).

20. C.-D. Schönwiese, *Klimatologie*, 5., überarbeitete und aktualisierte Aufl., Eugen Ulmer, Stuttgart, **2020**, pp. 346–355.

21. J. Marotzke, *Im Maschinenraum des neuen IPCC-Berichts*, Phys. Unserer Zeit, 53, **2022**, pp. 274–280.

22. C.-D. Schönwiese, *Klimatologie*, 5., überarbeitete und aktualisierte Aufl., Eugen Ulmer, Stuttgart, **2020**, pp. 362–367.

23. https://helmholtz-klima.de/aktuelles/welche-kipppunkte-erreichen-wir-bei-einhaltung-des-2-grad-ziels (last access: Oct 18, 2024).

24. M. Keil and B. P. Kremer (Eds.), *Wenn Monster munter werden*, 1st edn., Wiley-VCH, Weinheim, **2004**, pp. 145–150.

25. H. Pilcher, *Ab Nach Draußen*, 1st edn, Loewe Verlag, Bindlach, **2022**, pp. 110–111.

26. E. Breitmaier and G. Jung, *Organische Chemie*, 7., vollständig überarbeitete und erweiterte Aufl., Georg Thieme Verlag, Stuttgart, **2012**, p. 433.

3

In the Forest, in the Park, on the Hike

Abstract Before we get to experimenting and discovering, this chapter first describes the situation of the forests in Germany. The forest is in poor condition and a rethink or "reforestation" is urgently needed. Proposals for global climate rescue by 2050 are already available: Either plant 500 billion new trees or install 11 million new wind turbines. Also learn about new research on the topic of microplastics in the soil and in the air: Some trees are like vacuum cleaners and spider webs act as filters. With the help of tree lichens, you can determine the air quality of your environment. Lichens are fascinating creatures that can be discovered everywhere on hikes, travels, and on vacation. Some of them glow orange-red in UV light. How can you determine the height of trees? Quite simply: With a stick or a self-made mega triangle ruler. Further experiments are the secret of pine cones and the spore print of mushrooms.

© The Author(s), under exclusive license to Springer-Verlag GmbH, DE, part of Springer Nature 2024
A. Korn-Müller, *Scientific Secrets of Nature*,
https://doi.org/10.1007/978-3-662-69575-3_3

3.1 The Forest is Slowly Running Out of Steam

In Germany's forests, there are an estimated 90 billion trees, which pull about 52 million tons of carbon dioxide from the air each year, thus making a significant contribution to climate protection [1]. The most common tree species are spruces (26%), pines (23%), beeches (16%), and oaks (10%). Spruces are mainly found in the south and southwest of Germany in the low mountain ranges, while pines are mainly found in the northeast of the republic [1, 2]. The era of spruce monocultures is supposed to end, as they have nothing to counter climate change. The key to forest success is: replace conifers with mixed forests. This is already happening, the spruce population is declining, but trees grow very slowly. And it costs a lot of money. A new mixed forest devours 10,000–20,000 € per hectare (100 × 100 m) [2].

The importance of our forests is probably clear to everyone—we learn this in school in biology class. Forests are great for the climate, offer excellent recreation, and provide building materials for furniture and raw materials for paper production. Twenty years ago, our forests were dying from acid rain, today the enemies are deforestation, slash-and-burn, heat, drought, and bark beetles. Dead trees, like the ones found in large numbers in the Harz mountains or in southern Sweden, can no longer be used for furniture production, but only for burning and energy production, which in turn releases carbon dioxide (CO_2). However, forests should bind CO_2 and remove it from the atmosphere. A deciduous tree with a height of about 30 m and a crown diameter of about 15 m has more than 600,000 leaves, through which it absorbs about

9200 L (18 kg) of carbon dioxide and releases about 9100 L (13 kg) of oxygen per day in sunshine [3]. In addition, such a tree absorbs about 400 L of water through its roots every day and evaporates just as much through its leaves. The water vapor in the air, the humidity, therefore comes not only from rivers, lakes, and oceans, but here in Germany mainly from the 90 billion trees. Water vapor and the resulting clouds are important and cheap climate parameters.

The transformation of the forest is moving away from mono-spruce or -pine groves towards healthy and resilient mixed forests of at least three different tree species, such as oak, birch, sycamore, Douglas fir, pine, fir, and spruce, which cope better with drought, heat, and bark beetles [1, 2]. By the way, the trees and plants of a forest are not isolated "loners", but all are interconnected, with roots, fungi, microorganisms [4]. Similar to the landscape on the planet Pandora in the science fiction film "Avatar".

In 2019, a research team from Switzerland published a sensational study on forest area and climate neutrality. According to this, the entire world would be CO_2-neutral if an additional 9 million km^2 of the earth's surface were completely forested with 500 billion trees [5, 6]. This corresponds approximately to the land area of the USA or Canada or twice the area of the EU including Great Britain. However, comprehensive reforestation on this scale is problematic because the area must be suitable and the trees must grow for decades to extract enough CO_2 from the air [6].

In another interesting study by researchers from the Universities of Sussex and Aarhus in 2019, it was calculated that with 11 million new wind turbines with a total output of 52 terawatts, the energy needs of the entire world in 2050 could be covered [7]. The area required for this is around 5 million km^2, an area as large as the EU including Great Britain plus the North Sea. Taking into account the required minimum distances, there would be two onshore and three offshore wind turbines per km^2. Now that would be a real climate strategy!

Background

Trees—the world's vacuum cleaners

Microplastics not only pollute the entire oceans, but are present in even greater quantities in our soils. The size of microparticles is between 1 μm (1 thousandth of a millimeter) and 5 mm. All particles that are even smaller are called nanoplastics (15–1000 nm = 0.015–1 μm). Recent research shows that silver birches are capable of absorbing and storing microplastics from the soil, along with heavy metals and other industrial pollutants, and even breaking them down through their own bacteria [8]. The tree as a vacuum cleaner for contaminated soils. Brilliant! In addition, spruces, sessile oaks, and birches absorb nanoplastics through their roots and the water conduction paths in the trunk, branches, and leaves. However, the cleaning action of these trees unfortunately also has a downside. Buds, leaves, and bark serve as food for animals, which thus disperse and distribute the nanoplastic particles.

3.2 Spiderman Catches the Criminals—Spiders Catch Microplastics—Huh? Really Now?

Researchers at the University of Oldenburg found out in 2022 that spider webs unintentionally trap microplastics from the air. They examined spider webs in and around the illuminated shelters at bus stops [9]. As is well known, there are plenty of spiders there. On lightly trafficked roads, the scientists mainly detected PET (polyethylene terephthalate, 36%), which probably comes from textile fibers, in addition to six other types of plastic. On heavily trafficked roads, two candidates were mainly "struggling" in the net. Tire wear (41%) and PVC (polyvinyl chloride, 12%), which probably comes from the road markings. With this knowledge, sources of microplastics can be located in the future and even temporal progressions can be depicted.

Accelerating, braking, and cornering of motor vehicles generate more than 1 million tons of microplastics in the form of tire wear in Europe every year. This corresponds to about half of the total microplastic emissions in road traffic, about 2000 times more than the microparticles from the exhaust [10]. A British company founded in 2020 has invented a particle collection container that is mounted behind each wheel on the mudguard [10]. Each container contains electrostatically charged copper plates that catch the rubber particles. About 60% of the tire wear is filtered out of the air with this device. Great invention! The best part: The collected rubber particles can be upcycled and used to manufacture shoe soles, dyes, or new tires. Hopefully, these microplastic collectors will soon be available on the market as a series.

3.3   Survivors for Millions of Years: The Lichens

Lichens can be found not only on trees in forests and parks, but also in the mountains, on stones, on the coasts, and even in cities. Lichens are two organisms in one, consisting of a fungus (= mycobiont), in which green algae (= photobiont) have settled (sometimes also cyanobacteria) [11, 12]. They form a symbiosis: The green algae, due to their chlorophyll pigment, are able to convert sunlight into chemical energy, namely nutrients in the form of car-bohydrates, which benefit the fungus. In return, the fun-gus protects its algae, which are embedded in the upper fungal network, from excessive UV radiation and from drying out. In addition, it provides the algae with miner-als and water. A win-win situation. Since lichens do not have roots, they absorb nutrients, nitrogen, and moisture directly from the air, thereby purifying it. Lichens are not parasites, not freeloaders, but simply inhabitants of a cer-tain surface, called substrate. In Scandinavia, they serve as a food source for reindeer, as well as snails and deer. Microscopically small mites and tardigrades feel very com-fortable in lichens. The fungus is large and determines the shape, the algae is small and adapts.

There are three types of lichens: foliose/leaf lichens, fruticose lichens, and crustose lichens [11, 12]. They are found all over the world and are not deterred by extremes. Lichens grow in the icy poles, in hot deserts, in the trop-ics, in high mountains, everywhere. They are mainly found on solitary trees on bark and branches, on damp soil, on rocks and stones, but also on garden fences and on side-walks. You have certainly seen crustose lichens on your sidewalk at home, which look like flattened, old chew-ing gums. Lichens are often found on trees with suffi-cient nutrient supply, for example near fields, but also very often on street trees, where dogs do their small and large "business". Good fertilizer for the tree, good condi-tions for the lichen. In Fig. 3.1. you can see some types of lichens that I discovered on various hikes on vacation and on trips. The southern Black Forest in particular is very rich in lichens.

Lichens are true survival artists and can withstand temperatures between -200 °C and +80 °C. They grow extremely slowly at just under 1 mm per year, but can live up to 3000 years. A lichen about the size of a palm is estimated to be around 100 years old. Around 25,000 species of lichens are known worldwide, in Germany the number is estimated at about 1700 [11, 12]. Austria, with its alpine landscapes, has the highest lichen diversity. Lichens can appear in different shapes and colors and have been colonizing our earth for more than 420 million years, as evidenced by fossils.

Background

Lichens

In ancient Egypt, mummies were found whose bodies were, among other things, "stuffed" with fruticose and beard lichens. Particularly with the elk antler lichen *(Pseudevernia furfuracea),* also called tree moss, although the name "moss" is misleading and incorrect. This lichen emits a fungal-like scent. Fruticose lichens such as oak moss *(Evernia prunastri)* were used in the past for perfume production due to their bitter fragrances. Icelandic moss/Island moss *(Cetraria islandica)* is also not a moss, but a fruticose lichen, which was and is used in medicine as an expectorant in the form of cough teas and lozenges. Fox or wolf killer *(Vulpicida pinastri)* is one of the few poisonous lichens, which was hidden in baits in Scandinavia to kill foxes.

Some lichens, like the elegant sunburst lichen *(Xanthoria elegans),* have made it as research objects all the way to the ISS in space and survived there for a year [13].

On the websites of NABU (Nature Conservation Association Germany) [14] as well as BLAM (Bryological-Lichenological Working Group for Central Europe e.V.) [15] you will find numerous information about lichens. Bryology = moss science, lichenology = lichen science.

Fig. 3.1 Lichens, discovered on hikes and trips. **a** Fruticose lichen: ▶ Oakmoss/Plum lichen *(Evernia prunastri)* on an oak branch, Saxon Switzerland/Elbe Sandstone Mountains, size: 3 cm. **b** Fruticose lichen: Trumpet cup lichen *(Cladonia fimbriata)* on rotten wood, southern Black Forest, size of the cups: approx. 1.5 cm. **c** Foliose/Jelly lichen with cyanobacteria (blue-green algae): Flat-fruited shield lichen *(Peltigera horizontalis)* on a mossy granite rock, southern Black Forest, size: 26 cm. **d** Foliose lichen: Common yellow lichen *(Xanthoria parietina)* and Delicate scaly lichen *(Physcia tenella)* on a stone, Baltic Sea, stone size: approx. 1 m. **e** Crustose lichen: White blister lichen *(Phlyctis argena)* on sycamore maple, southern Black Forest, size: approx. 3 cm (the circular whitish spot). **f** Foliose lichen: Grooved key lichen (*Parmelia sulcata*) on lime tree, southern Black Forest, total size: 12 cm. **g** Foliose lichen: Delicate scaly lichen (*Physcia tenella*) on an old garden gate, Berlin-Lichterfelde. **h** Crustose lichen: Wall lichen *(Lecanora muralis)* on sidewalk, size: 1–4 cm. **i** Foliose lichen: Delicate scaly lichen *(Physcia tenella)* on the trunk of a summer lime tree, Great Garden, Dresden, size of the area: approx. 1 m. **j** Crustose lichen: Rock sulfur lichen *(Chrysothrix chlorina)* on sandstone rock, Kirnitzschtal, Saxon Switzerland/Elbe Sandstone Mountains. **k** Foliose lichen: Common yellow lichen (*Xanthoria parietina*) on lime tree trunk, length: 9 cm, and circular crustose lichens: Olive-green black cup lichen (*Lecidella elaeochroma*), diameter: 1 cm, Kurpark Bad Schandau/Saxon Switzerland. **l, m** Crustose lichen: Orange sea lichen *(Caloplaca marina),* on coastal rocks at the sea fjord Firth of Forth in Edinburgh/Scotland

Fig. 3.1 (continued)

3.3.1 Lichen Early Warning System: The Bioindicator for Clean Air

Lichens are very sensitive biosystems that react particularly to pollutants in the air. Sulfur dioxide (SO_2) is such a harmful gas, which is mainly released during the combustion of fossil fuels such as petroleum, natural gas and coal and by combustion engines and is referred to as chemical smog. Sulfur dioxide reacts with water to form sulfurous acid and finally with ozone in the troposphere to form

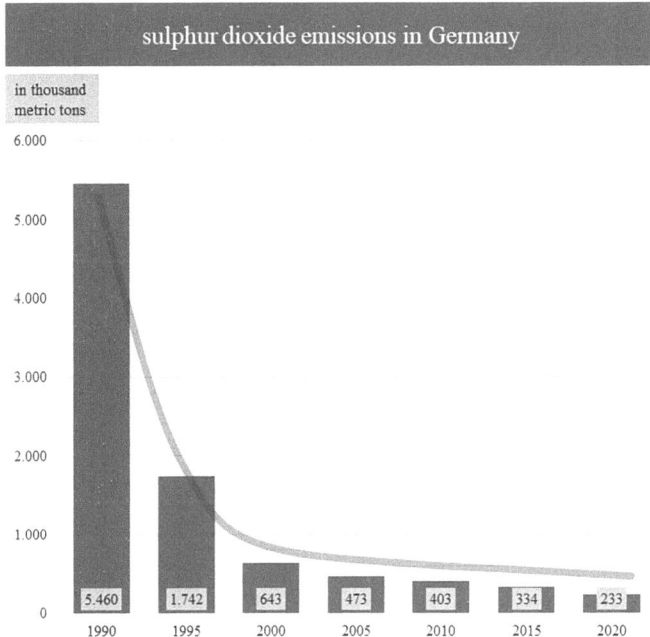

Fig. 3.2 Amount of sulfur dioxide emissions in Germany from 1990 to 2020 (Data: Federal Environment Agency [17]). Graphic: Melvin Müller

sulfuric acid and is washed down to the earth as "acid rain". The acidification of soils and waters to about pH 4–4.5 has dramatic consequences: decline of lake plankton, damage to amphibian and fish populations, leaching of important metals from the soil, release of toxic heavy metals from the soils and forest dieback [16].

Only through the massive use of flue gas desulfurization systems in power plants was the emission significantly reduced, as Fig. 3.2 shows. The toxic sulfur dioxide is converted with lime to gypsum (calcium sulfate), which is used in the construction industry. Consequently, sulfur dioxide emissions have been declining for decades, from 5.5 million tons in 1990 to 233,000 tons in 2020. That's

24 times less than in 1990 [17]. Exhaust gas catalysts in cars with combustion engines have also contributed to cleaner air. Harmful nitrogen oxides NO_x, especially nitrogen dioxide (NO_2), which reacts with water to form nitric acid, are reduced to harmless nitrogen on the surface of precious metals such as platinum, rhodium or palladium. Good for the lichens, good for the forest, good for the climate and good for us.

3.3.2 Experiment: Determining Air Quality with Lichen Counting

Since even the smallest amounts of sulfur dioxide and other harmful gases disrupt the growth of lichens, they serve as an ideal early warning system. One can say: The cleaner the air, the more lichens thrive. Nowadays, the air pollution from sulfur dioxide and nitrogen oxides has significantly decreased in cities. Therefore, even in large cities, lichens are increasingly found on trees, walls and sidewalks.

In fact, lichens are used to determine the quality of the air and are counted accordingly [18]. This is referred to as the so-called "lichen mapping to determine air pollution". As always in Germany, there is also a guideline for this, the VDI guideline (Association of German Engineers) VDI 3957 sheet 20 from 2017 [19]: "Biological measurement methods for determining and assessing the effects of air

pollutants (biomonitoring)—Mapping of lichens to determine the effects of local climate changes". Factors influ-
encing the lichens include precipitation, geological soil
conditions and geographical location.

Your Lichen Counting Device

Before you can determine the air quality in your environment yourself, you must first make a grid (50 × 20 cm)
with which you can count the lichens [18]. The counting
device consists of three 50 cm long bamboo sticks, which
span a grid of a total of ten fields, each of which is 10 ×
10 cm in size.

You will need

- 3 Medium-thick bamboo sticks (hardware store)
- Cord/package string
- Saw
- Drill
- Some skill
- Possibly a helping hand

How it works

First, shorten the bamboo sticks to about 60 cm, so about
10 cm longer than the actual lichen grid. This has two reasons: First, you then do not have to drill holes at the two
outermost ends—this does not work. Second, you have
enough space to set and drill the six holes with a distance
of 10 cm on the bamboo sticks and still have length left
over to comfortably grip the sticks with your hand. Six
holes are drilled into all three 60 cm long bamboo sticks at
the same height, each 10 cm apart (Fig. 3.3).

Now thread a piece of cord about 30–40 cm long
through the holes of one level and knot the string *before
and after* each hole. Otherwise, the bamboo sticks will slip.

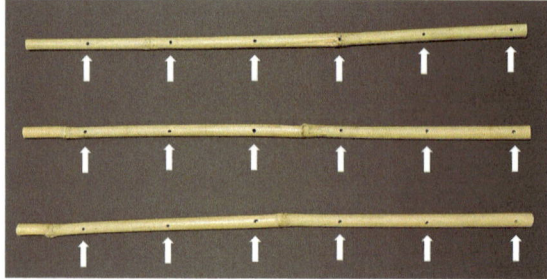

Fig. 3.3 Three bamboo sticks each with six drill holes at the same height, marked by white arrows. The distance of the holes on each bamboo stick is 10 cm. The distance between the three bamboo sticks is also 10 cm each. The top stick looks a bit crooked at the right end. I laid the sticks so that you can see the holes. When turning the sticks by 90 °, the holes face each other exactly and span a field with exactly 10 cm distance (see Fig. 3.4)

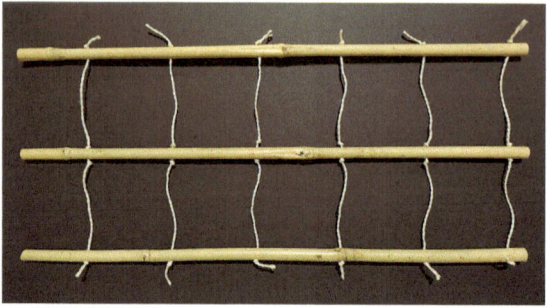

Fig. 3.4 Finished lichen grid. When pulled tight, there are 10 fields each with a 10 × 10 cm area (according to [18]). Now you can get started!

Make sure that the distances of each field are 10 × 10 cm. It's a bit of a tedious fiddling, but with a little patience and a helping hand, you can do it. When you have threaded and knotted the cords through all 6 holes, your finished lichen measuring device should look something like Fig. 3.4.

3.3.3   Experiment: How Good is the Air in Your Nearest Forest, Park or Residential Area?

You will need

- Your lichen grid
- 1–2 Tension rubbers with hooks or two helping hands

How it works
To get reasonably meaningful values, you should follow these rules:

- Choose medium-sized trees that are not too young (too narrow) and not too old (too wide).
- Officially, 3–15 trees are counted in an area from 250x250 m to 4x4 km. For us hobby air determiners, two to five trees every 100–300 m are sufficient.
- Only count lichens on trees from the same tree group, which is determined based on the bark texture into three groups:

  – Group 1: Pedunculate Oak, Wild Cherry, Silver Birch, and Black Alder
  – Group 2: Sycamore, Pear tree, Black Locust, Small-leaved Lime, Large-leaved Lime, Sessile Oak
  – Group 3: Norway Maple, Common Ash, Walnut tree, Apple tree, Poplar, Field Elm

Off to the forest or park for a healthy walk! You can conveniently carry your lichen grid and a rubber band rolled up in your backpack. Choose either only limes (group 2) or only pedunculate oaks (group 1); both types of trees are easy to recognize and are widespread. Maple and chestnut trees are also quite suitable. In my experience, lichens particularly like to grow on limes and sycamore. Proceed as follows: Choose the tree, unroll the grid and fix it at least 1 m above the ground on the tree trunk with a rubber band or by holding a companion (child, friend…). The grid is positioned at the most lichen-covered spot on the tree. Please make sure that the lichens are not damaged or destroyed by the lichen grid. If there are lichens within the fields, count as follows:

The maximum number of lichens per 10-by-10-cm field is set at 10. If the lichens on a tree fill the entire square, it is rated as 10. So a maximum of 10 x 10 = 100 counters are possible. Lichens often do not grow as individual "islands", but are interconnected as a network, so that individual counting is sometimes difficult or even impossible. In that case, simply estimate the number of lichens. The four trees in Fig. 3.5, which I photographed in spring or winter, serve as a "pattern" to show that the season does not play a major role in lichen counting.

Strictly speaking, one would have to distinguish and count the number of lichen*species*—but this is much too difficult for laypeople and hobby lichen seekers. With this simplified method, you do not get representative results in the strict scientific sense, but at least you get a rough idea of the air quality.

The Evaluation

Place the grid over the lichens of a tree and count the individual lichens in each of the ten squares. Only larger lichens with a diameter of about 1 cm are considered, the smaller ones are ignored [18]. In the four examples listed

Fig. 3.5   a Sycamore with high lichen density, **b** Sycamore with medium lichen density, **c** Summer linden with lichen carpet, **d** Summer linden without lichens. The ten fields of the grid are each numbered top left and the number of lichens is listed bottom right with a white border. The photos were taken in winter and spring

here, the following approximate lichen numbers would result:

Number of lichens on the tree in Fig. 3.5a: 10, 7, 10, 8, 8, 7, 7, 6, 5, 6. The sum gives the value 74 for the first tree. Now move on to the next trees and determine the lichen value for each tree. The tree in Fig. 3.5b has 44 lichens on its bark according to my count. On the tree in Fig. 3.5c, there is a whole carpet of lichens, making individual counting practically impossible. Here, I estimated the percentage spread, divided the sum by ten, and arrived at a value of 71. The fourth tree in Fig. 3.5d has no lichens at all.

In the end, you get four numerical values (depending on how many trees you have examined) and calculate the average from them: Add all numbers and divide by the number of trees. In our example: 74 + 44 + 71 + 0 = 189, divided by 4 equals an average = 47.3. This gives you the average number of lichens, which corresponds to the air quality value for the selected forest area, the so-called "lichen station". In Table 3.1 you can then read off the corresponding air pollution of the lichen station you have examined. 47.3 is just below 50 and corresponds to low air pollution and thus top air quality.

Table 3.1 Evaluation of the lichen count and determined air pollution or quality. (After Lit. [18])

determined air quality value	air pollution	air quality
>50	very low	Awesome! Take a deep breath!
37.5 - 50	low	Great!
25 - 37.4	moderate	Okay
12 - 24.9	high	Close windows!
0 - 12.4	very high	Aargh! *Throws up*

Fig. 3.6 Lichen grid on crustose lichens (wall lichen, *Lecanora muralis*) on a curb edge of a sidewalk

You can also determine the air quality in your street. Due to the increasingly clean air over the years, lichens thrive on many street trees, avenues, gardens, etc. Crustose lichens on sidewalks, pavements, or walls are also suitable for counting, as shown in Fig. 3.6. Look for lichens and determine the air quality!

By the way: Lichens usually grow on isolated trees in parks or on the roadside on the north or northeast side of the tree trunk to avoid direct sunlight. This way, the sensitive lichens avoid drying out. There is still enough light for the algae in the lichens to photosynthesize. Because of their preferred north orientation, lichens serve as an excellent compass—a natural navigation system.

3.4 Traffic Light on the Branch—Red Glow of the Yellow Lichens

With a bit of searching, you can find broken, overgrown branches in the undergrowth that are covered with all sorts of greenery. Moss and lichens have spread. Figure 3.7 shows such a branch that I discovered in a forest park.

The moss is easily recognizable by its shape and color (dark green). The white-green leaf lichen is the delicate scaly lichen *(Physcia tenella)* [20]. If you're lucky, the widespread common yellow lichen *(Xanthoria parietina)* is also on the branch. It thrives on nutrient-rich barks, often on street trees, but also on stones [21]. You have already discovered a true traffic light: green is there, yellow is present, and where is the red? All you need is a UV flashlight.

Fig. 3.7 A branch covered with moss (dark green) and leaf lichens (white green and yellow green) on the floor of a forest park

3.4.1 Experiment: Red Yellow Green on Branches

You will need

- UV flashlight (optimally with a wavelength of 365 nm)
- Branch or tree trunk with yellow lichens

How it works

To find a suitable branch with yellow lichens, I recommend that you search the forest, the park, the bushes, the undergrowth during the day for branches or tree trunks with yellow lichens. You will quickly find something. Either take the branch home or remember where it is.

Then go back to the forest or park to this branch with the yellow lichens at dusk or in the dark. When irradiated with UV light, this lichen glows in a spectacular orange-red color (Fig. 3.8). This so-called fluorescence is best seen in absolute darkness without scattered light.

Fig. 3.8   a Branch with yellow lichens *(Xanthoria parietina),* leaf lichens *(Physcia tenella)* and moss in daylight or white flashlight light. **b** The same branch illuminated with UV light (365 nm). The yellow lichens fluoresce with an orange-red color. Diameter of the largest yellow lichen on the branch: approx. 2 cm

What's the science behind it?

This orange-red glow is a fluorescence phenomenon. The lichen fungus contains the yellow pigment parietin (pronounced: Pa-ri-e-tin). This gives the common yellow lichen *(Xanthoria parietina)* its appearance and its name [22]. Parietin has it in itself. When irradiated with UV light, it fluoresces in a beautiful orange-red with an emission maximum at 610 nm. The yellow lichen thus operates a kind of "red-light district".

As already mentioned in Sect. 3.3, lichens contain green algae in their upper layer, which naturally contain chlorophyll. When irradiated with UV light, there is also red fluorescence, but this is so weak that it is simply outshone by the fluorescence of the parietin. The moss and the other type of lichen in Fig. 3.8 do not fluoresce (visibly) and therefore do not glow in the dark. Probably the excitation energy is not strong enough (light source too far away) or the plant surface does not let UV light through.

Due to the good air quality in cities and parks nowadays, you often find yellow lichens on street and park trees, even on stones. Take a closer look and always take your UV flashlight with you in the dark in the evening. Yellow lichens can be found everywhere, glowing in bright orange under UV light. An exciting yellow lichen hunt! Fig. 3.9 shows a small selection of yellow lichens discovered at home and on vacation. On the left half, the yellow lichens are illuminated with white light, on the right side the same spot is shown with UV light in the dark.

Tip: With the shorter, more energy-rich wavelength of 365 nanometers (nm), the orange-red glow comes into its own much better than with UV lamps with a wavelength of 395 nm.

You can find more spectacular fluorescence phenomena in nature in chapters 5and 6.

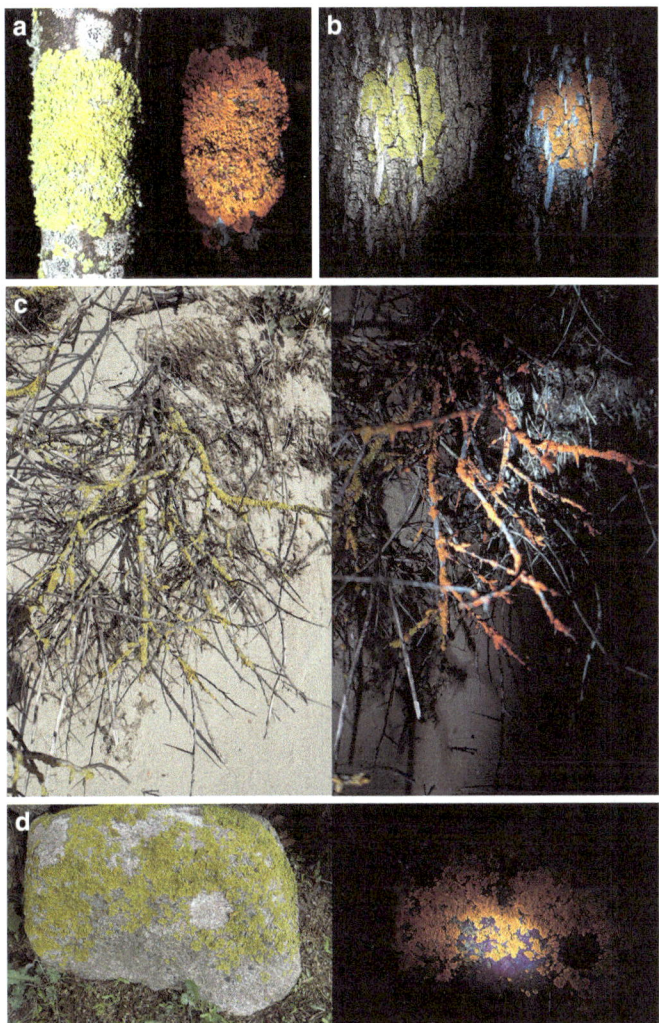

Fig. 3.9 Common yellow lichens on trees, shrubs and stones at home and abroad. On the left: white light (daylight, LED lamp), on the right: UV light (365 nm). **a** Linden in the spa park in Bad Schandau, Saxon Switzerland, length of yellow lichen: 9 cm; **b** Black poplar on the roadside (Belgium), size of yellow lichen: approx. 20 cm; **c** Thorn bush in the dunes of the Belgian North Sea, size of the bush: approx. 1 x 1 m; **d** Stone near the beach on the Baltic Sea (Fischland-Darß), width of the stone: approx. 1.40 m

3.5   Waldtraut is the Highest

Do you actually know where Germany's tallest tree can be found? In the southern Black Forest near Freiburg. If you ever vacation there or are nearby, it's best to start from the Waldhaus in Wonnhalde. By the way, you can learn a lot about the forest there, excursions and lectures are offered—I can only recommend it. From there, a sign-posted hiking trail leads over 4 km towards Schauinsland to the destination. Rather inconspicuous in a hollow among other trees, there stands a mighty Douglas fir about 68 m high. Since it does not stand alone, majestically like on a throne, one is somewhat disappointed. You could easily overlook it if a wooden sign did not alert hikers after about an hour's walk that they have reached their destination. The tree is apparently a female and bears the wonderful name "Waldtraut vom Mühlwald" on Illenberg (Fig. 3.10). There is even a wooden lounger for the exhausted hiker to relax and look up into the high treetops. The last measurement dates back to 2019 and was 67.10 m. A sign, which is located on the "local mountain" behind the forest house, indicates that this Douglas fir grows about 30 cm in height per year. So, in 2022 it would be about 68 m high. How can you actually measure or estimate the height of a tree with simple means? For this, a little math and only a few tools are necessary.

Fig. 3.10 Information board next to Germany's tallest tree: "Waldtraut vom Mühlberg"

3.5.1 Experiment: Tree Height Measurement using the Intercept Theorem—The Stick Method

You will need

- Stick, which is a bit longer than your extended arm (shoulder to hand)
- Folding rule or tape measure
- If available: Laser measuring device (hardware store)

How it works

Find a straight stick in the forest. You can also take a bamboo stick or lantern wood stick from home. It is important that the stick is longer than your extended arm so that you can comfortably hold it while measuring. Now adjust the stick to your arm length. To do this, take the stick in your hand (fist) and tilt it onto your extended arm. Now move the stick towards your shoulder until the end of the stick reaches the middle of your shoulder. The correct length goes from the fist to the shoulder. Grasp the stick again and hold it upright. Now the stick is "calibrated" and from the fist to the top tip exactly as long as your extended arm. Now you can start measuring trees as illustrated in Fig. 3.11.

Choose a tree of your choice. If you are on site, then preferably the "Waldtraut vom Mühlberg", because with

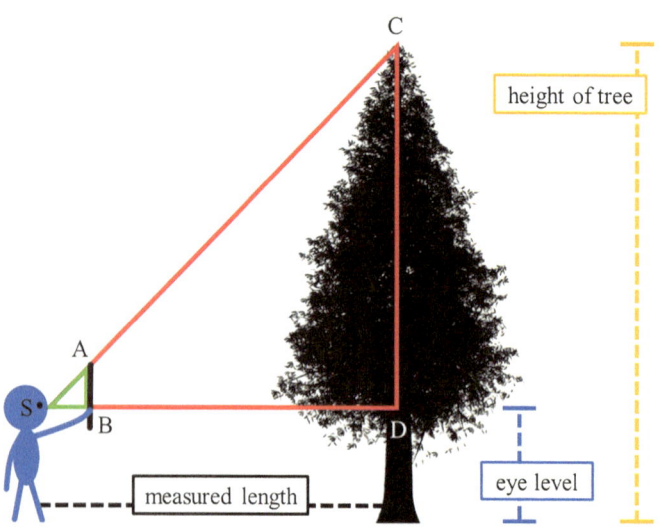

Fig. 3.11 Height measurement using the stick method based on the 2nd intercept theorem. Graphic: Melvin Müller

this specimen you know the presumed height and can first check how accurate your own measurement is. If not, that's fine too. Extend your arm with the stick in your fist and hold it as vertically as possible. Now walk backwards with a view to the tree to be measured and aim over the stick towards the tree. Observe the tip of the stick at about 45° and move away from the tree until the tip of the stick coincides with the tip of the tree. It's best to squint one eye. The whole thing is a bit shaky, depending on how high or low the stick is held. If the tip of the stick and the treetop coincide, stay put. You now have to measure the distance from here to the tree. The easiest way is by pacing. The number of steps times the stride length gives the length. You can determine your stride length with a folding rule (from toe to toe). For me, it's about 80 cm. You can also determine the distance to the tree very easily and quite accurately with a laser measuring device.

Finally, you need to add your eye level to the determined length to get the final result of your measurement. Simply measure the eye level from the ground with the folding rule. For me, it is 1.75 m.

What's the science behind it?

This method is based on the second intercept theorem, which states that the ratio of the parallels is equal to the ratio of the sections on one of the rays. The stick held in the hand is parallel to the tree trunk—these are the two parallels. The corresponding rays form the lines of sight from the eye towards the top of the tree and the tree trunk (parallel to the ground). If we look at Fig. 3.11, then the following applies:

$$\overline{SB}/\overline{AB} = \overline{SD}/\overline{CD} \qquad (3.1)$$

Sought: $\overline{CD}$= Tree height (at eye level).

Given: $\overline{AB}$= Stick length, $\overline{SB}$= Arm length = Stick length.

Now solve Eq. (3.1) for $\overline{CD}$:

$$\overline{CD}/\overline{SD} = \overline{AB}/\overline{SB} \qquad (3.2)$$

$$\overline{CD} = \overline{AB}/\overline{SB} \times \overline{SD} \qquad (3.3)$$

Since $\overline{AB}$ is the same size as $\overline{SB}$ they cancel each other out.

$$\overline{CD} = \overline{SD} \qquad (3.4)$$

$\overline{SD}$ = is the distance from the standpoint to the tree in meters.

$\overline{SD}$ corresponds to the distance $\overline{CD}$ thus the height of the tree at eye level.

The total height of the tree results from $\overline{SD}$ plus eye level.

Examples and Evaluation

In my measurement of the "Waldtraut vom Mühlberg" (approx. 68 m, 2022), I walked the distance several times and on average needed 81 steps. With a stride length of 0.80 m plus 1.75 m eye level, I arrive at a total height of around 67 m (exact: 66.55 m), thus slightly lower than the "official" height of 68 m. The error rate was therefore $\pm 2\%$. For error estimation, I determined the height of a lantern in a park using a laser measuring device (6 m) and arrived at 6.55 m with the stick method ($\pm 9\%$ deviation). I determined the base height of a monument to be 8.95 m, its exact height was 8.50 m. The measurement inaccuracy was around $\pm 5\%$. At low heights, the error deviation of 2–9% has a stronger effect than with very tall trees.

3.5.2 Experiment: Tree Height Measurement with Mega Triangle Ruler

You will need

- Styrofoam or hard foam board (hardware store)
- Cutter or jigsaw

How it works

A hard foam board from the hardware store (e.g. type XPS 035, 125 x 62.5 x 2 cm, hard foam 200-G, Fa. Recit, Hornbach hardware store, price: 3.50 €) is particularly suitable for this outdoor experiment for two reasons: It is very stable and absolutely waterproof. Cut a square of, for example, 50 x 50 cm from the board and halve it diagonally, as shown in Fig. 3.12. This results in

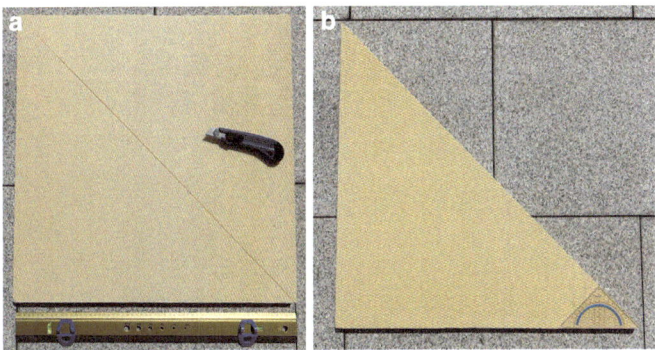

Fig. 3.12 **a** Square foam board cut diagonally into two right-angled, isosceles triangles. **b** The triangle ruler shows the 45° angle at the base

two right-angled and isosceles triangles with the two base angles of 45° each (Fig. 3.12b), essentially an oversized set square.

Choose a tree of your choice and place the triangle on one of the two legs. Now move it with a view to the tree to be measured and remove it as far from the tree until the vanishing point of the base (the long side) of the triangle coincides with the top of the tree. It's best to squint one eye and press your head flat on the ground. Lying down works reasonably well, but you shouldn't be wearing your nicest clothes. Children will be delighted to finally be allowed to wallow in the dirt—in the service of science (Fig. 3.13).

To find the exact line of sight, you would strictly have to position your head below the base angle—in a hole, a trench, a hollow. We'll leave that and take an error value of approx. +10% determined by me for this inaccuracy. If the line of sight and tree top coincide, you have to measure the distance from the triangle to the tree, add 10% and you have the approximate height of the tree or other object.

Fig. 3.13 Height measurement using the mega triangle ruler

Alternatively, you can hold the mega triangle ruler at eye level like the stick method and then bring the vanishing point into line with the top of the tree. The tree height is then calculated from the distance from the triangle to the tree plus the eye level. The measurement inaccuracy here is like the stick method at 5–9% (6.55 m instead of 6 m and 8.95 m instead of 8.50 m).

What's the science behind it?
With this method, you use the mathematical fact that the two legs of the triangle are the same length (Fig. 3.14). Once you have determined the length of one of the two legs, you automatically know the length of the other leg (= height of the tree) [23].

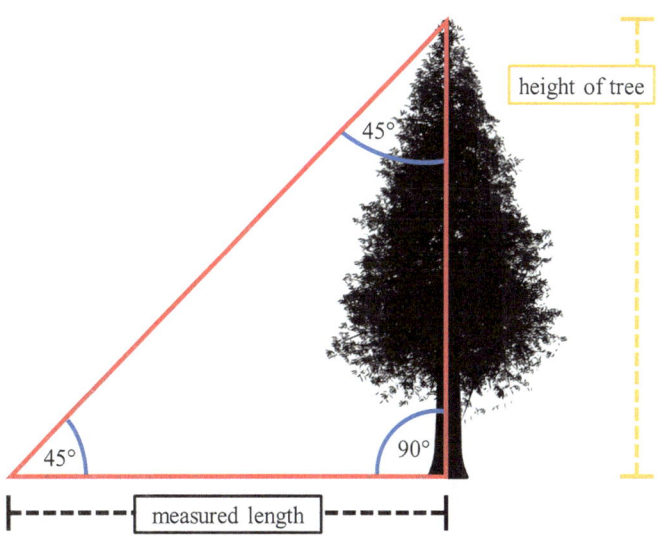

Fig. 3.14 Height measurement using the mega triangle ruler (Graphic: Melvin Müller)

3.6   Magic Cone

If you are ever hiking or walking through a forest or park, please collect some pine cones from the ground. Cones from other conifers are also welcome, but they are rarer and pine cones are nice and small and handy. Cones seem to be alive and want to play a trick on us—with a water sprayer and a jigsaw you can get to the bottom of the mystery of the closed and opened cones.

3.6.1   Experiment: Cone Drawbridge

You will need

- Some pine cones
- Water sprayer (flower sprayer)
- Jigsaw
- Some patience

How it works

The experiment is most impressive when you start with opened cones. If you have closed cones in front of you, then let them lie in the warm apartment for a while. After a few hours, the pine cones should have opened. Now the experiment can begin. Put an opened cone in the refrigerator or freezer for a few hours. The cone remains open, even if it "squatted" in the cold overnight. Now spray an opened pine cone with water. Make it properly wet. And lo and behold: After one to two hours, the cone is closed. If you let it dry in the warm room again, the cone opens again. You can repeat this opening and closing experiment as often as you like.

What's the science behind it?
As Goethe so beautifully puts it in his "Faust", *"Two souls, alas, dwell in my breast, one wants to separate from the other."* In the cell walls of the cones, two "souls" also dwell in the form of lignin and cellulose [24]. Lignin is a large biopolymer bridged over oxygen atoms from phenol units. Cellulose consists of thousands of glucose molecules that form long chains or fibers. These two wood substances are responsible for the opening and closing of the cone scales. In humidity or rain, the scales close to protect the seed leaves. In warm weather and dryness, the scales open again and release the winged seeds to be blown into the surrounding area by the wind [24]. Cold or temperature has no influence on the behavior of the cones.

To be able to examine the interior more closely, I sawed an open pine cone from top to bottom with a jigsaw (Fig. 3.15). This is admittedly a bit tedious but then you can see these two "souls" very nicely.

As with a bimetal, the two wood layers lignin and cellulose are connected to each other. In contrast to lignin, cellulose can absorb a lot of water and swell. The cellulose fibers are aligned in such a way that they curl the cone scales inward when they swell [24]. The lignin layer is pulled up like a drawbridge until the cone is "tight". The same thing happens with a bimetal, but due to a temperature change.

When it gets dry, the water in the cellulose evaporates and the curvature gradually subsides—the drawbridge goes down again. After a few hours, the scales are fully extended and you can start the experiment anew. Material researchers are trying to implement this physical mechanism of swelling and shrinking of a material as a motor,

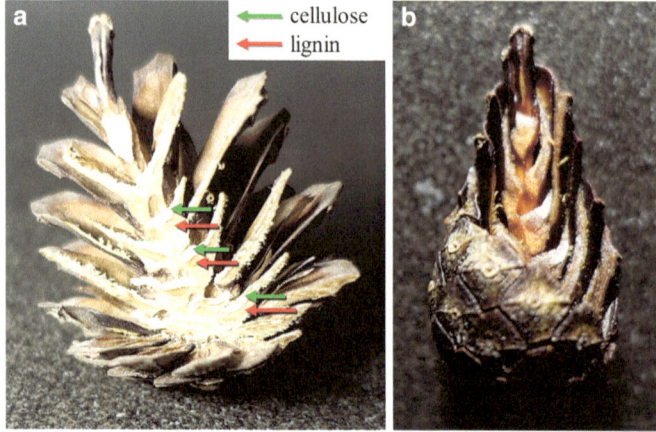

Fig. 3.15 **a** shows a sawn open, opened pine cone and **b** the same cone after spraying with water. The white layer consists of swellable cellulose (green arrows), the brownish layer of lignin (red arrows)

for example to open and close doors, windows or flaps [24]. Another implementation idea is a power-free humidity detector for wooden houses. Nature has already shown how it works. Bravo, cones!

3.7   Autumn is Mushroom Time

If you collect mushrooms yourself in the autumn and are well versed in it, then you can carry out a nice and very simple experiment with (edible) gilled mushrooms [25]. But it also works well with purchased champignons, which are even offered in supermarkets all year round.

3.7.1 Experiment: Mushroom Spore Print

You will need

- Some non-toxic, edible gilled mushrooms (e.g., champignons)
- Sheet of white paper
- Thick books or large stones

How it works

First, remove the stem of the mushroom. Then place the mushroom cap on a sheet of paper and weigh it down with a thick, heavy book or a heavy stone. This will slightly flatten the mushroom, as shown in Fig. 3.16a.

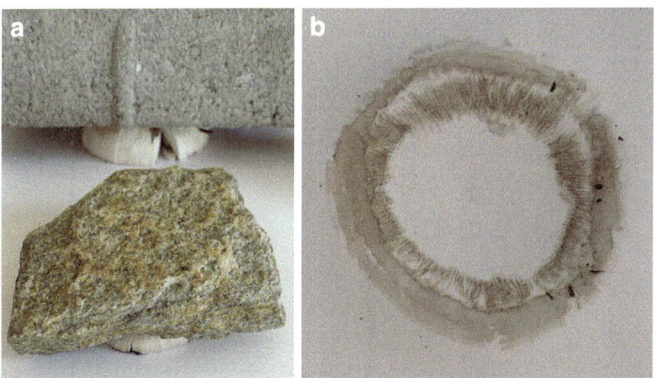

Fig. 3.16 a Champignon caps on a sheet of paper—weighed down with a stone. **b** After a day of pressing, the spore print appears on the sheet of paper

After a day (overnight), you can remove the stone/book and admire a beautiful gill print on the sheet (Fig. 3.16b).

What's the science behind it?
The heavy weight of the stone presses the gills of the mushroom cap firmly onto the sheet of paper. The mushroom spores contained in the gills are transferred to the paper and remain as a visible, brownish "trace".

References

1. a) https://www.bundeswaldinventur.de/dritte-bundeswald-inventur-2012/klimaschuetzer-wald-weiterhin-kohlenstoff-senke/ (last access: Oct 18, 2024). b) https://www.wald.de/waldwissen/wie-viel-kohlendioxid-co2-speichert-der-wald-bzw-ein-baum/ (last access: Oct 18, 2024). c) https://www.sdw.de/ueber-den-wald/waldwissen/wald-in-zahlen/ (last access: Oct 18, 2024).
2. T. Baier, Hilfe dringend gesucht, Süddeutsche Zeitung, 18.01.2022, p. 2.
3. A. Goldberg (Hrsg.), Biosphäre, Bd. 6 Gymnasium Sachsen, 1. Aufl. 2. Druck, Cornelsen Verlag, Berlin, 2021, pp. 110–112.
4. M. Groß, Das Mikrobiom der Pflanzen, Nachr. Chem. 70, 2022, pp. 79–80.
5. J.-F. Bastin, Y. Finegold, C. Garcia et al., The global tree restoration potential, Science, 365, 2019, pp. 76–79.
6. J. W. Veldman, J. C. Aleman, S. T. Alvarado et al., Comment on "The global tree restoration potential", Science, 366, 2019, pp. 318–320.
7. P. Enevoldsen, F.-H. Permien, I. Bakhtaoui et al., How much wind power potential does Europe have? Examining European wind power potential with an enhanced socio-technical atlas, Energy Policy, 332, 2019, pp. 1092–1100.

8. E. Leusmann, Mikro- und Nanoplastik, Nachr. Chem. 70, 2022, p. 73.
9. I. Goßmann, R. Süßmutth and B. M. Scholz-Böttcher, Plastic in the air?! – Spider webs as spatial and temporal mirror for microplastics including tire wear particles in urban air, Sci. Total Environ., 832, 2022, 155008.
10. S. Roeder, Auf Teilchenfang, Dresdner Neueste Nachrichten, 24./25.12.2022.
11. V. Wirth und U. Kirschbaum, Flechten einfach bestimmen, 2., aktualisierte Aufl., Quelle & Meyer Verlag, Wiebelsheim, 2017, pp. 7–30.
12. L. Urry, M. Cain, S. Wasserman, P. Minorsky und J. Reece, Campbell Biologie, 11., aktualisierte Aufl., Pearson Deutschland, München, 2019, pp. 886–887.
13. https://www.123pilzsuche-2.de/daten/details/ZierlicheGelbflechte.htm (last access: Oct 18, 2024).
14. NABU – Naturschutzbund Deutschland e. V. https://www.nabu.de/tiere-und-pflanzen/pilze-und-flechten/14125.html (last access: Oct 18, 2024).
15. Bryologisch-Lichenologische Arbeitsgemeinschaft für Mitteleuropa e. V. https://blam-bl.de/blam/blam-verein.html (last access: Oct 18, 2024).
16. https://www.umweltbundesamt.de/themen/luft/luftschadstoffe-im-ueberblick/schwefeldioxid (last access: Oct 18, 2024).
17. https://www.umweltbundesamt.de/daten/luft/luftschadstoff-emissionen-in-deutschland/schwefeldioxid-emissionen#entwicklung-seit-1990 (last access: Oct 18, 2024).
18. M. Keil und B. P. Kremer (Hrsg.), Wenn Monster munter werden, 1. Aufl., Wiley-VCH, Weinheim, 2004, pp. 133–144.
19. https://www.vdi.de/richtlinien/details/vdi-3957-blatt-20-biologische-messverfahren-zur-ermittlung-und-beurteilung-der-wirkung-von-luftverunreinigungen-biomonitoring-kartierung-von-flechten-zur-ermittlung-der-wirkung-von-lokalen-klimaveraenderungen (last access: Oct 18, 2024).

20. V. Wirth und U. Kirschbaum, Flechten einfach bestimmen, 2., aktualisierte Aufl., Quelle & Meyer Verlag, Wiebelsheim, 2017, p. 47.
21. V. Wirth und U. Kirschbaum, Flechten einfach bestimmen, 2., aktualisierte Aufl., Quelle & Meyer Verlag, Wiebelsheim, 2017, p. 35.
22. https://www.biologie-seite.de/Biologie/Gew%C3%B6hnliche_Gelbflechte (last access: Oct 18, 2024).
23. H. Pilcher, Ab Nach Draußen, 1. Aufl., Loewe Verlag, Bindlach, 2022, pp. 124–125.
24. C. Zollfrank, Holzbasierte Aktuations-Systeme, FG Biogene Polymere, Wissenschaftszentrum Weihenstephan für Ernährung, Landnutzung und Umwelt, TU München, 2015. https://www.holz.tum.de/fileadmin/w00bqw/holz/Bilder/Aktuell/Cordt_Zollfrank_Vortrag.pdf (last access: Oct 18, 2024).
25. H. Pilcher, Ab Nach Draußen, 1. Aufl., Loewe Verlag, Bindlach, 2022, pp. 174–175.

4

At the Pond, Puddle, or Lake

.

Abstract You can't imagine what's going on in a single drop of water from a pond or puddle! It's teeming with life, an invisible hustle and bustle, and the best part is: you can observe it yourself—without a microscope—using the "Laser-Drop-Method" with a laser pointer and a syringe from the hardware store. Microorganisms, green algae, ciliates, copepods, oral mucosa cells, and red blood cells. Bustling videos for scanning included. How to make light refraction and reflection visible? This can be done with ordinary soap bubbles that turn into disco balls. For the cold season, frozen lakes offer a real "experimental playground". With a laser pointer, ice floes turn into light organs and the ice thickness of a body of water can also

Supplementary Information The electronic version of this chapter contains additional material, which can be accessed via the following link https://doi. org/10.1007/978-3-662-69575-3_4. The videos can be played by clicking on the DOI link in the legend of a corresponding figure, or by scanning this link with the SN More Media App.

A. Korn-Müller, *Scientific Secrets of Nature*,
https://doi.org/10.1007/978-3-662-69575-3_4

be measured with a laser pointer. The classic must not be missing: The sounds on a frozen lake. Why this is so, you will find out in this chapter.

4.1 Making Microorganisms Visible with a Laser Pointer—Without a Microscope

Tens of thousands of different microorganisms, animal and plant plankton, such as green algae or ciliates, lead a life invisible to our naked eye in the waters of the earth. With diameter sizes between 10 and 500 μm, they remain hidden from us (1 μm = 10^{-6} m = 1 thousandth of a mm). This changed with the development of powerful microscopes, with which humans are able to penetrate and explore ever smaller spheres of the microcosm. From bacteria, viruses, proteins to individual atoms [1]. However, there is also a surprisingly simple method to make microscopically small microorganisms visible without the help of a microscope. All you need is a green or red laser pointer and a plastic syringe.

4.1.1 Experiment: Creating Detailed Images with the "Laser-Drop Method"

If you send the laser beam of a conventional laser pointer through a water droplet containing cells or microorganisms, you can make them visible as an enlarged shadow image. Since the intensity of the red light is not as strong as that of the green laser beam, the images with the red laser pointer do not appear quite as bright. However, this proves to be advantageous for photographing and filming.

You will need

- Green or red laser pointer with a wavelength of 532 nm (green) or 650 nm (red) and a power between 1 and 10 mW (reference: Internet approx. 5–20 €) ⚠
- Plastic syringe (approx. 20 mL volume, hardware store or pharmacy)
- 2 Tripods with clamp (laboratory or photo tripods)
- Cup

How it works

The liquid to be examined is drawn into the syringe and the syringe is clamped vertically onto a tripod. The laser pointer is also fixed horizontally with a clamp to the second tripod. The distance between the laser pointer and the syringe does not play a major role, about 20–30 cm (Fig. 4.1) have proven to be effective. Any room, house or garage wall, a brick wall or a real canvas can serve as a "screen". The distance between the syringe and the "screen"

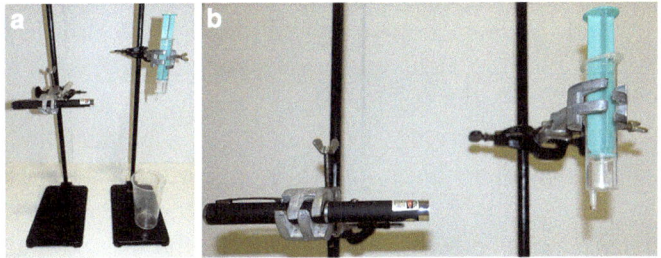

Fig. 4.1 **a** Experimental setup of the laser-droplet method: Laser pointer and syringe. **b** Detail

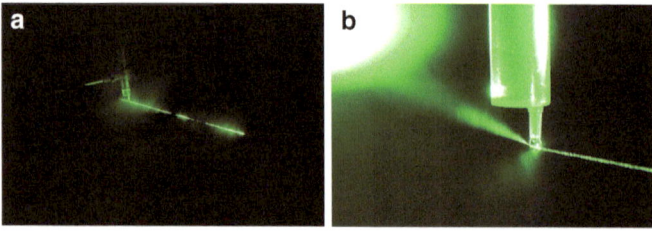

Fig. 4.2 Laser beam of a green laser pointer (532 nm), made visible with theater fog. **a** The beam runs from right to left and just below the drop. **b** Close-up of the laser beam as it "shoots" through the hanging drop

should be about 2 m in order to see (and measure) the cells well. Figure 4.1 shows the general experimental setup.

Just before the experiment, gently squeeze a drop from the syringe until the drop just barely hangs on the syringe opening. It often happens that several drops fall down until finally a drop remains hanging. Therefore, a cup should be placed under the syringe to catch the liquid. Now align the laser beam as horizontally as possible to the plane exactly on the drop. You can slowly "feel your way" from top to bottom or vice versa. This requires a steady hand and some practice. Figure 4.2a shows the green laser beam from the source towards the wall ("screen") just below the drop hanging on the syringe tip. For the examination, the laser beam must therefore be aligned a little bit upwards so that it shoots exactly through the drop (Fig. 4.2b). The laser light was made visible using theater fog from a handheld fog machine. Caution! Please observe the safety instructions when handling laser beams: Do not shine such a laser beam into the eyes! Do not hold against reflective materials (mirrors, glasses, metal surfaces)! The laser light is reflected and refracted by the drop and also by the plastic of the syringe—please protect your eyes, for example with sunglasses.

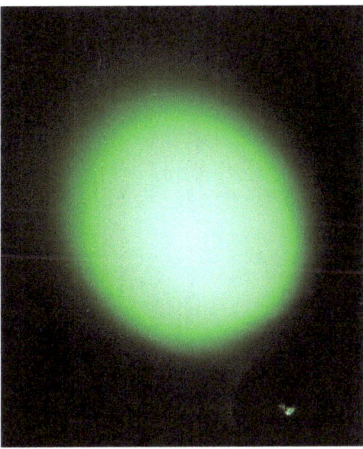

Fig. 4.3 Negative test with tap water. The light of a green laser pointer penetrates a water droplet, but does not show any shadows of cells or microorganisms

For a "negative test", I drew tap water into the syringe and irradiated several drops with the laser pointer. As Fig. 4.3 shows, no cells or microorganisms are present.

When everything is ready: Lights out—spectacle on!

4.1.2 Experiment: Oral Mucosal Cells

The simplest way to obtain usable cells is from oral mucosal cells, which are found in large numbers on the inner sides of the cheeks [2, 3]. One takes a sip of water and rinses the mouth by vigorously moving it back and forth. The cells rubbed off in this process are now in the water. Spit it into a glass, draw it up with a syringe—done.

If this is too "unappetizing" for you, there is a more elegant method: Using a cotton swab, you take a cell smear by rubbing the cotton head with moderate pressure on the inner sides of the cheeks (similar to the Corona test). The wet cotton head is washed out in about 10 mL of water by moving it back and forth and turning it, and the "mouthwash" is finally drawn up into the syringe. This method has the advantage of better hygiene and achieving a sufficiently high dilution. This way, you can easily recognize the individual cells. With the "spit method", you get a huge amount of cells, some of which overlap and cover each other.

Figure 4.4 shows the magnification of oral mucosal cells in red and green laser light. Their original size is about 40–80 µm with irregular, diffuse, sometimes rectangular,

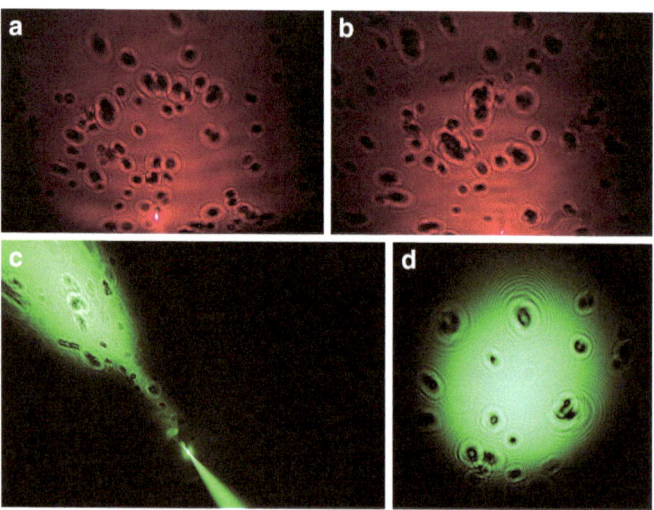

Fig. 4.4 Shadow images of oral mucosal cells. Distance from droplet to wall: 2 m, cell shadow size between 8 and 13 cm in diameter. **a** and **b** using a red laser pointer. **c** and **d** using a green laser beam

sometimes round shape with a round, dark cell nucleus in the middle [3, 4]. You often see different sizes and also some "fragments" created by mechanical abrasion.

4.1.3 Experiment: Microorganisms from the Pond, Puddle, Lake

"Arm" yourself with a plastic syringe and walk or cycle to a pond, puddle, pool, or lake. From the shore, draw about 20 mL of water into the syringe (it's best to "rinse" the syringe once with the water). Even if the water sample looks clear and pure from the outside, it is teeming with thousands of microorganisms, known as plankton. To date, around 40,000 plant and animal plankton species are known, which are found in all bodies of water on Earth [5]. There are over 7000 aquatic green algae worldwide, about 6300 of which are freshwater green algae [6]. They are widespread in lakes, ponds, and puddles and occur as eukaryotic single and multicellular organisms, but also form chains (filaments) and colonies. The typical size of green algae is 20–100 μm [7]. In addition to green, blue, yoke, and diatoms, there is a multitude of single-celled organisms, such as ciliates, as well as countless tiny animals, such as the copepod (oarfoot crab), in the water [8]. However, this experiment is to neglect the type of microorganisms, because a clear determination of the biological species is not possible with this method—but also not necessary. As a silhouette, you can see almost circular organisms as well as rod-shaped, filamentous specimens [8, 9]. The examined water samples come from a pond in the Great

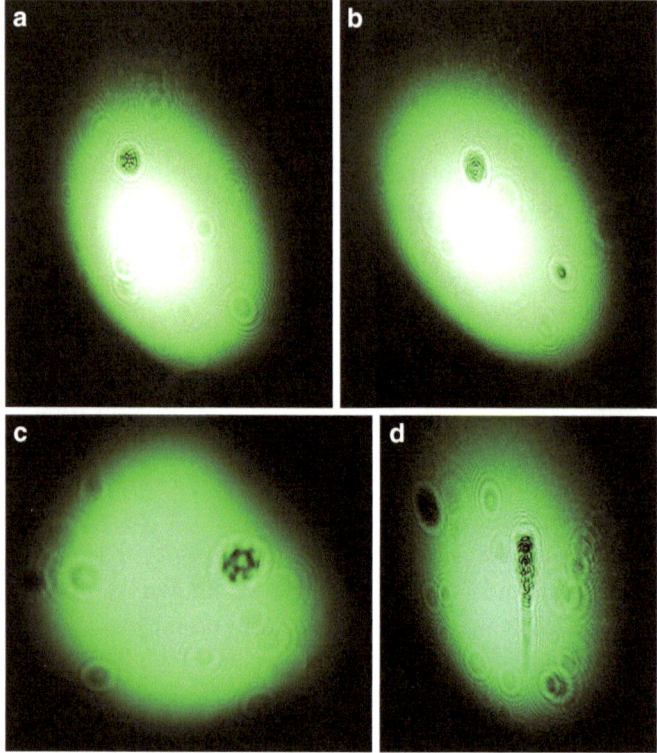

Fig. 4.5 Silhouettes of freshwater plankton in laser light. Distance from droplet to wall: 2 m, shadow size approx. 8–16 cm in diameter. **a–d** in spring. **e–h** in summer

Garden in Dresden, drawn once in spring (March, April) and once in summer (July, August). It is clearly recognizable that the reproduction of the plankton has massively increased by summer (Fig. 4.5). In Fig. 4.5a it could be a star algae *(Micrasterias)* (average original size: 40–300 μm) [10, 11]. The Fig. 4.5e–g show wildly swimming microorganisms that look like small balls of wool and race across the "screen". Presumably, these are the nimble ciliates

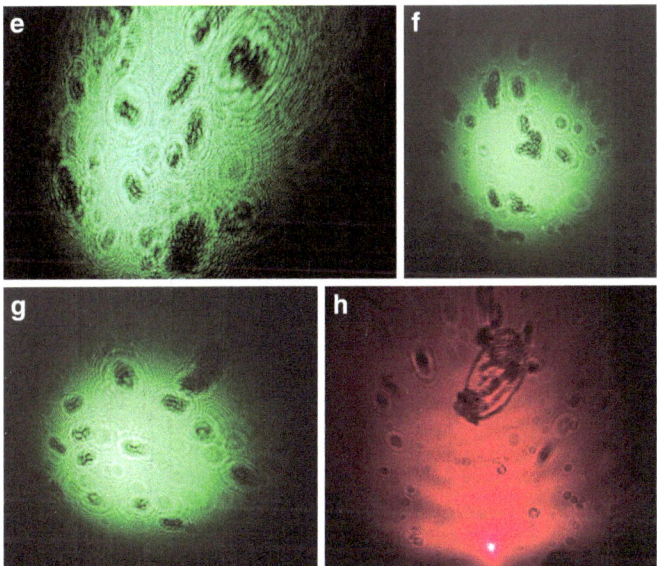

Fig. 4.5   (continued)

(average original size: approx. 20–200 µm) [12] as well as rotifers (original size: 40–3000 µm) [13].

To see what's going on in a single drop of pond water in high summer, please scan the URL in Fig. 4.6. I recorded the video in green laser light and it shows hundreds of microorganisms frantically moving around.

In numerous samples of summery pond water, all hell breaks loose! Creatures that move quickly and hectically, some of them rowing, dart through the picture as if stung by a tarantula, constantly changing their direction. It's unbelievable! A swarm like no other! These are, among others, copepods (*Copepoda,* original size approx. 0.5–2 mm) [14, 15] and presumably ciliates and rotifers [12, 13]. The drifting organisms are probably green algae. Figure 4.7 shows a still image of a copepod (oarfoot crab) from a video.

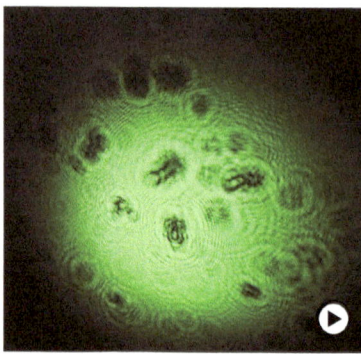

Fig. 4.6 The video shows various microorganisms in summer pond water as enlarged shadows using the "laser drop method" with a green laser pointer. Music: Caribbean World by Mezaproduction Aleksandr B. Karabanov (pixabay) URL: ▸ https://doi.org/10.1007/000-a6a

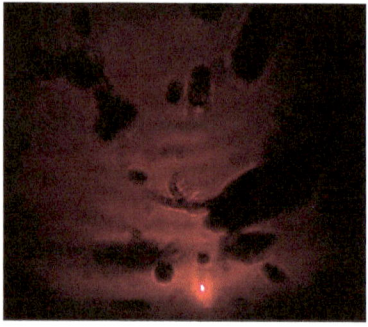

Fig. 4.7 Silhouette of a copepod (oarfoot crab) in red laser light as a still image from a video. Original size approx. 0.5–2 mm. Other types of plankton are also visible. Distance from droplet to wall: 2 m

A spectacular video with wildly darting copepods (oarfoot crabs) can be accessed under Fig. 4.8. One observes a wide range of different microorganisms of all sizes and shapes, which is not surprising, as in the water sample

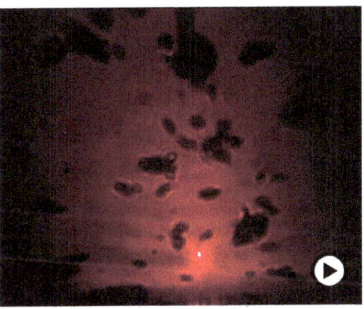

Fig. 4.8 Hectic swarm of various microorganisms in summery pond water. The video shows, among others, copepods and ciliates as enlarged shadows with the "laser drop method" using a red laser pointer. Music: Caribbean World by Mezaproduction Aleksandr B. Karabanov (pixabay) URL: ▸ https://doi.org/10.1007/000-a69

"young and old" as well as many different species live together [8].

4.1.4 Experiment: Moss Extract

In the fall, collect about a tablespoon of moss and soak it in a bowl of water overnight. The next day, gently squeeze out the green moss and draw the "moss water" into the syringe. The moss extract contains a multitude of cells, cell fragments, tiny and huge moss fibers. A real "mixed bag". The "laser drop method" makes the invisible visible. In Fig. 4.9 you can see the result.

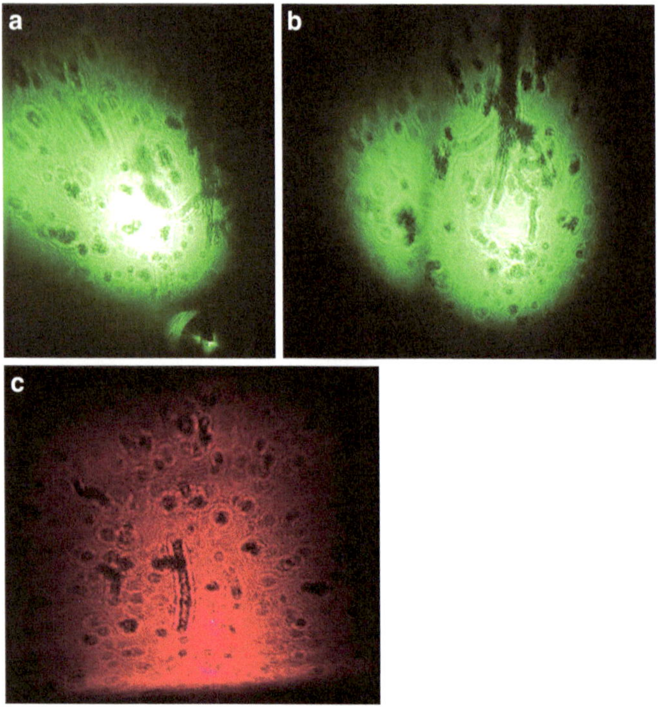

Fig. 4.9   a–c Moss extract under "laser fire". The largest shadows have a length of up to 90 cm. This corresponds to an actual size of about 0.5 mm. Distance from droplet to wall: 2 m

4.1.5   Experiment: Red Blood Cells (Erythrocytes)

With their 6–8 µm diameter, red blood cells (erythrocytes, short: erys) represent the absolute limit of the "laser drop method". For a blood sample, my oldest daughter had to volunteer. She pricked her fingertip with a sterile

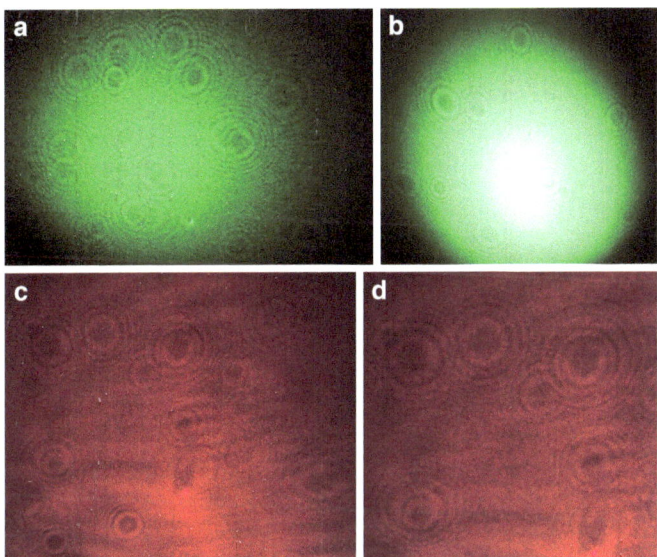

Fig. 4.10 Image of red blood cells (erythrocytes) using the "laser drop method". **a** and **b** with a green laser pointer. **c** with a red laser beam. **d** Close-up of **c**. Distance drop to wall: 2 m, cell shadow size approx. 1.5–2 cm in diameter

cannula to squeeze out a small drop of blood. I took this tiny drop with a pipette and put it in 40 mL of water to get a sufficiently high dilution. In a 10- or 20-mL dilution, there are still far too many erythrocytes present, so that one simply sees a "gray mass" in the shadow image. A drop has a volume of about 50 µL (0.05 mL), a very small one about 25 µL and contains around 125 million red blood cells [16]. Diluted to 40 mL (40,000 µL), there are about 78,000 erys per drop. With about 200,000–500,000 cells per drop of blood, the white blood cells (leukocytes, short: leukos) are in the absolute minority [16] and play virtually no role in the diluted liquid with 125–625 specimens per drop, but were occasionally sighted. The photos in Fig. 4.10 show the typical "saucer" or donut shape of the red blood cells.

What's the science behind it?

For the calculation of the magnification effect, the water drop is assumed to be a sphere, it acts as a spherical lens. Incident light rays are partly refracted and reflected. From Huygens' principle, Snell's law of refraction results with the angles θ_1 and θ_2 (Eq. 4.1) [17, 18].

$$n_1 \sin \theta_1 = n_2 \sin \theta_2 \qquad (4.1)$$

Here, the refractive indices $n_1 = 1.0$ for air (medium 1) and $n_2 = 1.33$ for water (medium 2) apply. Light rays are refracted at the interface of the two media air and water (Fig. 4.11). In addition, according to Huygens' principle for reflection, the angle of incidence α is equal to the angle of reflection α' (Fig. 4.11).

The majority of the laser rays pass through the liquid drop and hit the cells. The magnification V through a

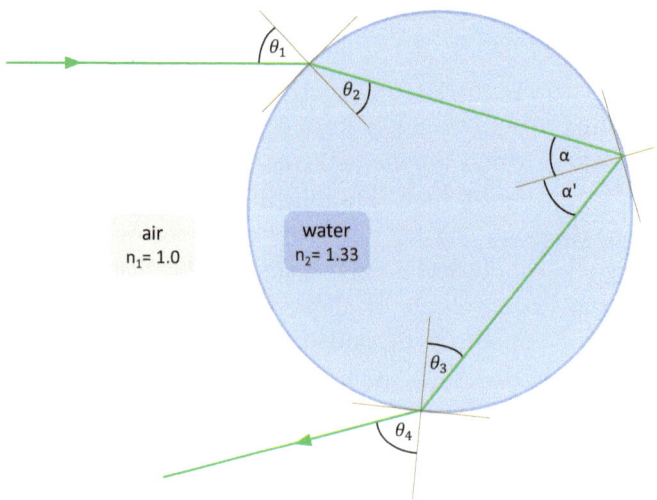

Fig. 4.11 A drop idealized and simplified as a sphere. Light rays are partially refracted and reflected at the spherical surface. Graphic: Melvin Müller

spherical refracting surface can be described with Eqs. 4.2 [17, 18]. Here, B stands for the image height and G for the object height.

$$V = \frac{B}{G} \tag{4.2}$$

The distance of the droplet to the image plane (image distance) is described with b. The object distance g corresponds to the radius of the droplet when the object is exactly at the center of the circle, which is assumed as an approximation. The refractive index of air is $n_1 = 1.0$ and that of water is $n_2 = 1.33$. Figure 4.12 shows the beam path with the magnification effect.

For very small angles, the approximation applies that $\sin \theta$ is approximately equal to the angle θ. Inserted into the law of refraction (Eq. 4.1) results in Eq. 4.3.

$$n_1\theta_1 = n_2\theta_2 \tag{4.3}$$

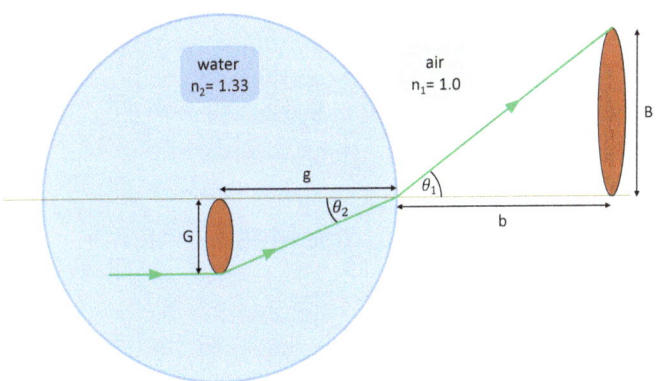

Fig. 4.12 Simplified representation of the beam path in the droplet as an idealized spherical surface and resulting magnification of a cell; g=radius droplet, b=distance droplet—image plane. Graphic: Melvin Müller

In the right-angled triangle (Fig. 4.12), the tangent of the angle θ corresponds to the length ratio of the magnitudes of opposite to adjacent and for axis-near, i.e., very small angles, the approximation $\tan \theta \approx \theta$ also applies here (Eqs. 4.4 and 4.5).

$$\tan \theta_1 \approx \theta_1 = \frac{B}{b} \tag{4.4}$$

$$\tan \theta_2 \approx \theta_2 = \frac{G}{g} \tag{4.5}$$

With Eq. 4.3 and 4.6 results by inserting and solving for V.

$$n_1 \frac{B}{b} = n_2 \frac{G}{g} \Rightarrow V = \frac{B}{G} = \frac{n_2 b}{n_1 g} \tag{4.6}$$

Using Eq. 4.5, the magnitude of the magnification can be calculated. The quotient of n_2/n_1 is 1.33. The droplet is approximated as a sphere and with a usual diameter of about 3 mm, the radius $g = 1.5$ mm is assumed [19].

The concentric rings around the cells are diffraction and interference patterns caused by the coherent laser light and slightly distort the actual shape and size of the object.

4.1.6 Evaluation of the Measurements

To verify the calculated magnification factors, the cells projected onto the wall were measured with a tape measure. The image distance, or the distance b from the droplet to the screen, was optionally set to 1 m or 2 m. Since the cells or organisms in the droplet are constantly moving due to convection forces or self-excitation, a "quick hand"

Table 4.1 Laser droplet measurement of the approximate size of cells and microorganisms

	typical diameter	magnification factor & expected size	measured size (diameter)
oral mucosa cells	40-80 μm	1,773 (2 m distance) 7-14 cm	8-13 cm
		887 (1 m distance) 3.5-7 cm	4-8 cm
microorganisms in the pond (plankton, green algae)	40-100 μm	1,773 (2 m distance) 7-18 cm	8-16 cm
		887 (1 m distance) 3.5-9 cm	3-7 cm
moss water (algae and fibres)	various sizes and shapes, 30 μm to 500 μm – 0.5 mm	1,773 (2 m distance) 5-89 cm	5-18 cm 27 cm 84 cm 90 cm
blood cells (erythrocytes)	6-8 μm	1,773 (2 m distance) 1.0-1.4 cm	1.5-2 cm

is needed for the measurement. However, for a rough estimate, it does not matter down to the millimeter. In Table 4.1, the measurement results are summarized in an overview.

The original size of the diameter of adult oral mucosa cells is 40–80 μm [3, 4]. With $b = 1$ m and $g =$ droplet radius $= 1.5$ mm, Eq. 4.5 gives a magnification of $V = 887$. Measured diameters of the cell shadows on the wall were usually between 4 and 8 cm. The theoretical value for a cell with a diameter of 40 μm is 3.5 cm, with 80 μm about 7 cm. With $b = 2$ m, a magnification of $V = 1773$ is obtained. Measured diameters of the cell shadows on the wall were usually around 8–13 cm, some showed lengths of 15–17 cm. The theoretical value for a cell with a diameter of 40 μm is about 7 cm, and with 80 μm about 14 cm.

The original size of the diameter of green algae (phytoplankton) is on average about 40–100 μm [7, 8]. With b = 1 m, a magnification of $V = 887$ is obtained. Measured diameters of the cell shadows on the wall were about 3–7 cm. The theoretical value for a cell with a diameter of 40 μm is about 3.5 cm, with 100 μm diameter about 9 cm. With b = 2 m, a magnification of $V = 1773$ is obtained. Measured diameters of the cell shadows on the wall were often between 8–9 cm and about 12–16 cm. The theoretical value for a green alga with a diameter of 40 μm is about 7 cm, with 100 μm diameter about 18 cm.

Red blood cells (erythrocytes) are very small with their 6–8 μm diameter [16]. With b = 2 m, a magnification of $V = 1773$ is obtained. Measured diameters of the cell shadows on the wall were between 1.5 and 2 cm. The theoretical value for a cell with a diameter of 6 μm is about 1 cm, with 8 μm about 1.4 cm. White blood cells (leukocytes) are about 5–20 μm in size, but due to their negligible number, they do not play a role in this experiment, but they were recognizable.

4.1.7 Conclusion and Evaluation

The measured sizes agree quite well with the theoretical values and are within the calculated range. Especially with the blood, the measured values were a little too high. However, the measurements should only be seen as a rough approximation, as there are several sources of error to consider with this method. Not all cells are exactly in the middle of the droplet, they are distributed everywhere and move due to flow forces (convection) or self-motion in the hanging droplet. The radius of the droplet can vary depending on the size of the droplet at the syringe tip. Depending on the alignment of the laser beam, distorted or elongated images on the wall result. The measurement inaccuracy when measuring the shadows also plays a role. Therefore, the "droplet-laser method" does not achieve exact science, but this is completely sufficient within the framework of such a simple and yet so astonishing experiment for easy replication. It is amazing that one can actually see the living microorganisms and cells—even if only as gray-black shadows. For students, it would certainly be a great experience in the classroom if the board is teeming with microorganisms, copepods, or mucosal cells of fellow students (or the teacher). So: Let it teem at home—or if you are a biology teacher—in the classroom on the wall! This also wonderfully combines biology and physics.

Finally, some safety aspects

Never shine laser light into the eyes! Do not beam against glass, foils, mirrors, or reflective surfaces! Protect the eyes best with tinted safety glasses/sunglasses!

4.1.8 Experiment: Disco Ball Made of Soap Bubbles

The previous experiment dealt with optical effects in a droplet. With a simple experiment, the refraction or reflection of light on spherical surfaces can be made spectacularly visible. I first saw this experiment at the Science Show on soap bubbles by Mr. Joachim Lerch at the "Science Days" in Europa-Park, who filled huge soap bubbles with theater fog.

You will need

- Soap bubbles (1 tube e.g. Pustefix®)
- Smoke from a cigarette or fog from an electric vaporizer
- Laser pointer (red or green)
- Darkness

How it works

First, find a suitable darkness: basement, room, garage, outside in the evening, balcony in the evening, etc. I recommend laying a cardboard base on the floor because blowing soap bubbles can make a mess. Tip: Soap bubbles work best outside when it has rained. Then the humidity is very high and the bubbles last longer. In dry air, the water in the soap bubble evaporates quite quickly, so the shell becomes thinner and thinner until it finally bursts.

Blow some soap bubbles and catch *one* bubble with the blowing part again. Now carefully blow a small amount of cigarette smoke or the fog of an e-cigarette into the soap bubble. Often they break in the process, but please do not be discouraged. With a little patience, you can do it. There doesn't have to be much smoke or fog in the bubble—a hint of smoke is completely sufficient. Turn off the light or carry out the experiment in the dark. Now shine the laser pointer through the smoke-filled soap bubble. The laser beam becomes immediately visible because the smoke or fog particles reflect and scatter the light (Fig. 4.13).

If you send a red and a green laser beam through the soap bubble at the same time, the soap bubble mutates into a Christmas tree or disco ball illuminated from the inside (Fig. 4.14).

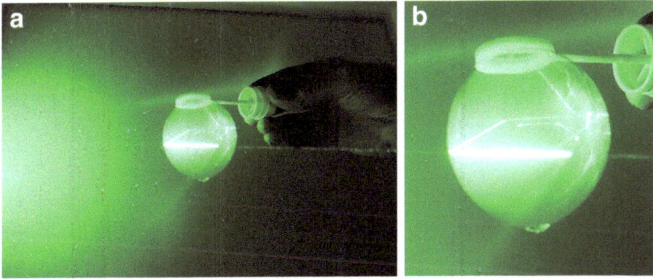

Fig. 4.13 **a** The laser beam of a green laser pointer shoots through a smoke-filled soap bubble. The beam goes from right to left and is refracted, reflected twice, and transmitted. **b** Detail

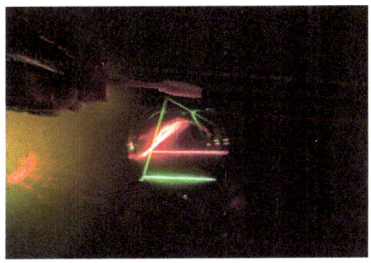

Fig. 4.14 Laser beams from a red and a green laser pointer shoot through a smoke-filled soap bubble. The beams go from right to left. The reflections inside the soap bubble are clearly visible and transform the bubble into a mini disco ball

Safety Aspects

Never shine laser light into the eyes! Do not aim at glass, foils, mirrors, or reflective surfaces! It is best to protect the eyes with tinted safety glasses/sunglasses!

What's the science behind it?

As Fig. 4.13 shows, both the partial refraction and the partial reflection can be clearly seen at the interface between air and soap bubble surface. Part of the light beam undergoes a double reflection within the bubble, while the majority of the laser light passes straight through the soap bubble, as can be seen from the bright spot on the wall at the back (Fig. 4.13a). With two differently colored laser beams, the optical effects appear even more spectacular (Fig. 4.14).

4.2   Ice Play in Winter

Is it freezing cold outside? Is the lake, the stream, the puddle frozen over? Then get out there and perform icy-beautiful experiments!

4.2.1 Experiment: Disco Ice Floe

Take a small ice floe and a laser pointer and the lights will dance.

You will need

- Small, not crystal clear ice floe (frozen puddle, frozen pond)
- Laser pointer (red or green)
- Darkness

How it works

Wait for darkness. Fortunately, it gets quite dark in winter from about 4:30 pm, so you don't have to stay up late with your children. Break a small ice floe from the edge of a frozen body of water. It should look somewhat "cloudy" and not completely clear. For my recordings, I used both ice floes from the rain barrel in the garden and pieces of ice from a frozen lake with a thickness of about 0.5-3 cm.

However, you could also make an ice floe in the freezer of the refrigerator with water in a bowl yourself and independently of non-existent cold. In the darkness, simply illuminate the piece of ice with a laser pointer. Instantly, the disco lights dance on the background! Move the laser beam back and forth so that it illuminates different parts of the ice plate. Choose a flat background as a "screen". In Fig. 4.15 you can see my photos with different pieces of ice/ice floes.

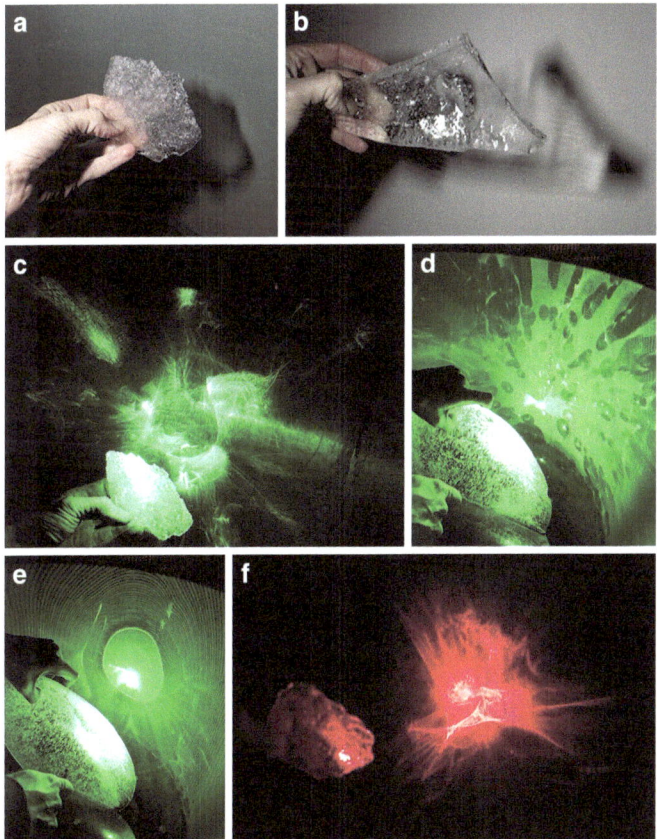

Fig. 4.15   Ice floe disco: **a** Thin ice floe 0.5 cm. **b** Thick ice floe 3 cm. **c–e** A small and a large ice floe are illuminated with a green laser pointer. **f** A piece of ice is irradiated with red laser light

Safety aspects

Never shine laser light into the eyes! Do not beam against glass, foils, mirrors or reflective surfaces! Protect the eyes best with tinted safety glasses/sunglasses!

What's the science behind it?

The more "opaque" the ice floe is, that is, the more gas inclusions and irregularities the ice crystal has, the more beautiful patterns result. The laser beam breaks at the countless gas bubbles, inclusions and cracks, is scattered and thus creates spectacular patterns. Two videos can be found under Fig. 4.16 (red laser pointer) and Fig. 4.17 (green laser pointer), in which I used two differently thick ice floes.

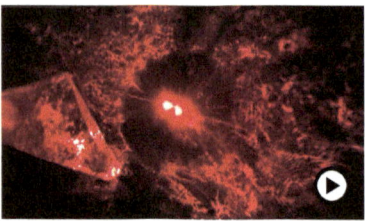

Fig. 4.16 Dancing, red disco lights with ice. The video shows the spectacular lighting effects when moving, red laser light shines through an ice floe. Music: Milky Way—Ambient Space Music by JuliusH (pixabay) URL: ▸ https://doi.org/10.1007/000-a68

Fig. 4.17 Dancing, green disco lights with ice. The video shows the spectacular lighting effects when moving, green laser light shines through an ice floe. Music: Please Calm My Mind by Oleksii Kaplunskyi/Lesfm (pixabay) URL: ▸ https://doi.org/10.1007/000-a6b

4.2.2 Experiment: Measuring Ice Thickness

When it has been really icy cold with 10–14 days of sub-zero temperatures and the standing waters are frozen solid, you can easily estimate the thickness of the ice using a laser pointer. According to the DLRG, you should only step onto the ice when it is at least 15 cm thick.

DLRG warning: Never go onto the ice alone and children should only be accompanied by an adult!

You will need

- Frozen lake, pond, pool, … body of water
- Laser pointer (red or green)
- Darkness

How it works

From the shore, in approaching darkness from late afternoon, shine a laser pointer as vertically as possible or at least at a 45° angle through the ice. You should choose a surface that shows the clear ice, so that the laser beam can pass unhindered to the bottom. The entry point of the beam is revealed by a small scattering around the beam. The focused laser light is not scattered in clear ice. Only when it exits under water does the beam become wider and brighter. This difference allows you to roughly estimate the thickness of the ice layer. Figure 4.18 shows "measurements" with a green and a red laser pointer as well as a close-up. The estimated ice thickness here is about 5 cm.

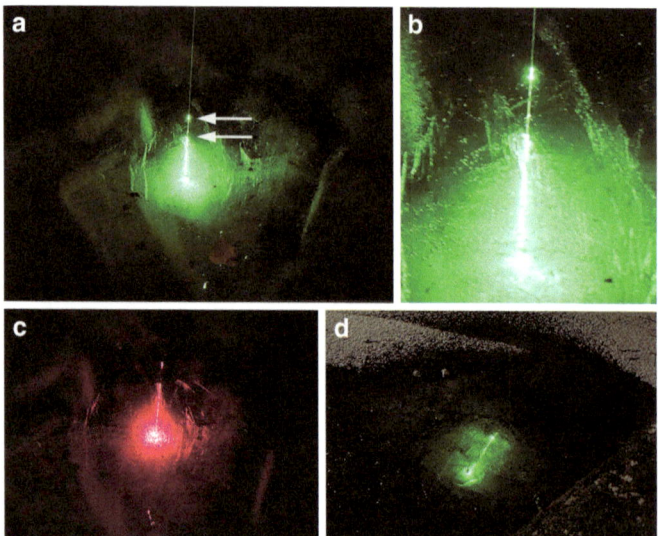

Fig. 4.18 Measurement of the ice thickness of a frozen pond with **a** a green and **c** red laser pointer. The clear area between the entry and exit point of the laser beam corresponds to the thickness of the ice. Here: approx. 5 cm. **b** Close-up of **a**. **d** Measurement on another winter day. Ice thickness: approx. 3 cm

Safety Aspects

Never shine laser light into eyes! Do not aim at glass, foils, mirrors or reflective surfaces! Protect your eyes with tinted safety glasses/sunglasses!

What's the science behind it?

In clear ice, the coherent laser beam is not scattered and passes unhindered through the crystal. However, when the laser light exits the ice under water, it hits tiny suspended particles, which cause scattering and thus the "lighting up" of the beam. It almost looks like a lightsaber. At the entry point on the ice surface, there is also a tiny scattering or refraction due to the different media of air and ice. If you look closely, two scattering points can be seen, because there is some water as a third medium between the air and the ice surface, which has formed due to milder temperatures.

4.2.3 Experiment: Bing-Kling-Titscher-Titscher-Dirr

Perhaps you already know this experiment from your own experience with your children or even as a child yourself? When you throw ice cubes or stones onto the surface of a frozen body of water, these typical "Ditscheredirr" sounds are produced, which sound like laser cannons from science fiction spaceships or lightsabers in Star Wars. Due to the low friction, the "projectiles" slide incredibly far. Even Christian Morgenstern (1871–1914) knew about this and described it very aptly in his children's poem "When it becomes winter" [24]:

"The lake has gotten a skin,
so that you can almost walk on it,
and if a big fish comes swimming,
it bumps its nose.

And if you take a pebble
and throw it on, it goes clink
and titscher – titscher – titscher – dirr …
Hey, you funny pebble!

It chirps like a little bird
and pretends to fly like a swallow -
but finally my pebble
lies far, far out on the lake."

You will need

- Frozen, standing water (lake, pond)
- (small and larger) Ice floes, ice cubes or stones

How it works

Break different sized pieces of ice/ice floes from the frozen shore of a lake. These flat floes zoom particularly far! If the ice is too thick, simply bring some ice cubes from home in a bag. At external minus degrees, these will not melt. It also works great with stones, as described by Christian Morgenstern in the quoted poem. At a shallow angle, you now hurl the throwing pieces "with gusto" onto the frozen ice surface. The parts zoom over the smooth surface and produce the characteristic sound. I still find this great even as an adult! The larger the lake, the more effective the "song". Throw several ice cubes onto the ice surface at the same time! The song becomes a small symphony: The Sound of (S)**i**(len)**ce**.

What's the science behind it?

The smooth surfaces of a frozen lake and a thrown ice floe experience only little frictional forces, so that the throwing pieces glide almost unbraked a long distance over the ice. Even rough stones bounce far out. During the sliding over the ice surface, audible sound waves in the high frequency range are generated, which we perceive as unusual and particularly sensitive [20]. Normally, the audible sound is a jumbled mix of a multitude of different sound waves or frequencies at all heights and depths [21]. This sound wave mix is transmitted by the air at around 330 m/s into our ear [21]. So far, so known. But: Through solid matter like ice, high frequencies, i.e. high tones, move faster than low tones [22]. As a result, the sound waves separate, the stronger, the longer the throwing piece is on the ice. The larger the frozen lake, the more impressive the effect. The high sound waves race away from the lower tones, so that the acoustic noise is pulled apart. Like in a bike race, where suddenly the fastest sprinters break away from the rest of the slower field and rush off. In physics, this effect is called acoustic dispersion [20]. Dispersion refers to the dependence of the propagation speed on the frequency of light or sound waves [22, 23]. The splitting of light in a prism or in raindrops into the rainbow colors is such a dispersion effect. Normally, the propagation speed of sound waves is dependent on the medium, but not on the frequency. But the "solid-fluid" surface ice behaves exceptionally dispersive because it can be stretched and compressed within certain limits [20]. By the way: Each body of water has—depending on the ice thickness—its individual pitch, which asserts itself as audible sound. The thinner the ice, the higher the tone. An ice cover of 10 cm and less is great to "hear". Thick ice over 15 cm sounds significantly lower and less dramatic [20].

Background

Speed of sound

- in air (20 °C): approx. 340 m/s
- in water (20 °C): approx. 1400 m/s
- in ice (−4 °C): approx. 3200 m/s
- in steel (20 °C): approx. 5100 m/s

References

1. M. T. Madigan, J. M. Martinko, D. A. Stahl und D. P. Clark, *Brock Mikrobiologie kompakt,* 13., aktualisierte Aufl., Pearson Verlag Deutschland, München, **2015**, pp. 13–30.
2. A. Korn-Müller und P. Eimer, *Was dein Körper alles kann*, Fischer Sauerländer Verlag, Frankfurt, **2016**, p. 34.
3. B. P. Kremer, *Mikroskopieren – Ganz Einfach*, 1. Aufl., Franckh-Kosmos Verlag, Stuttgart, **2021**, pp. 76–79.
4. http://www.rs-pfiffelbach.de/pdf/Aufgaben/Bio7_4.pdf (last access: Oct 18, 2024).
5. *Biosphäre*, Band 6 Gymnasium Sachsen, 11. Aufl. (Hrsg.: A. Goldberg), Cornelsen Verlag, Berlin, **2021**, pp. 154.
6. L. Urry, M. Cain, S. Wasserman, P. Minorsky und J. Reece, M., *Campbell Biologie*, 11. aktualisierte Aufl., Pearson Verlag Deutschland, München, **2019**, pp. 807.
7. M. T. Madigan, J. M. Martinko, D. A. Stahl und D. P. Clark, *Brock Mikrobiologie kompakt,* 13., aktualisierte Aufl., Pearson Verlag Deutschland, München, **2015**, pp. 549–552.
8. H. Streble, D. Krauter und A. Bäuerle, *Das Leben im Wassertropfen*, 13., überarbeitete Aufl., Franckh-Kosmos Verlag, Stuttgart, **2017**.
9. B. P. Kremer, *Mikroskopieren – Ganz Einfach*, 1. Aufl., Franckh-Kosmos Verlag, Stuttgart, **2021**, pp. 100–105.
10. H. Streble, D. Krauter und A. Bäuerle, *Das Leben im Wassertropfen*, 13., überarbeitete Aufl., Franckh-Kosmos Verlag, Stuttgart, **2017**, pp. 174–175.
11. B. P. Kremer, *Mikroskopieren – Ganz Einfach*, 1. Aufl., Franckh-Kosmos Verlag, Stuttgart, **2021**, p. 104.
12. H. Streble, D. Krauter und A. Bäuerle, *Das Leben im Wassertropfen*, 13., überarbeitete Aufl., Franckh-Kosmos Verlag, Stuttgart, **2017**, pp. 55–58 und 216–241.
13. H. Streble, D. Krauter und A. Bäuerle, *Das Leben im Wassertropfen*, 13., überarbeitete Aufl., Franckh-Kosmos Verlag, Stuttgart, **2017**, pp. 65–67 und 254–273.

14. V. Storch und U. Welsch, *Kurzes Lehrbuch der Zoologie*, 8., neu bearbeitete Aufl., Springer Spektrum Verlag, Berlin Heidelberg, **2012**, pp. 527–528.

15. H. Streble, D. Krauter und A. Bäuerle, *Das Leben im Wassertropfen*, 13., überarbeitete Aufl., Franckh-Kosmos Verlag, Stuttgart, **2017**, pp. 292–293.

16. K. Kunsch, *Der Mensch in Zahlen*, 1. Aufl., Gustav Fischer Verlag, Stuttgart, **1997**, pp. 46–47.

17. P. A. Tipler und G. Mosca, *Physik*, 8. Aufl. (Hrsg.: P. Kersten und J. Wagner), Springer Spektrum Verlag, Berlin, **2019**, pp. 1031–1071.

18. H. Bannwarth, B. P. Kremer und A. Schulz, *Basiswissen Physik, Chemie und Biochemie*, 4., aktualisierte Aufl., Springer Spektrum Verlag, Berlin, **2019**, pp. 127–132.

19. https://de.wikipedia.org/wiki/Tropfen (last access: Oct 18, 2024).

20. H. J. Schlichting, *Zwitschern auf dünnem Eis*, Spektrum der Wissenschaft, **2019**, 12, 72–73.

21. H. Bannwarth, B. P. Kremer und A. Schulz, *Basiswissen Physik, Chemie und Biochemie*, 4., aktualisierte Aufl., Springer Spektrum Verlag, Berlin, **2019**, pp. 62–64.

22. P. A. Tipler und G. Mosca, *Physik*, 8. Aufl. (Hrsg.: P. Kersten und J. Wagner), Springer Spektrum Verlag, Berlin, **2019**, pp. 521–522.

23. W. Demtröder, *Experimentalphysik 1*, 9. Aufl., Springer Spektrum Verlag, Berlin, **2021**, pp. 396–397.

24. Christian Morgenstern, *Wenn es Winter wird*, 3. Aufl., Hase und Igel Verlag, Garching b. München, **2013**.

5

Searching for Traces with the UV Lamp—A Light Spectacle

Abstract I would not have thought it possible, how much in nature glows or fluoresces so colorfully under UV light. You should definitely get a UV flashlight and then shine it around outside in nature or on vacation. You will experience your colorful wonder! Shine it in the garden, in the forest, in the park, illuminate vegetables, plants, mosses, lichens, mushrooms, stones, minerals, spiders, and woodlice. A glowing vegetable smiley or a creepy pepper? No problem with UV light! Even garbage can be tracked down in nature, and LED lights glow magically. What the beaches of the North and Baltic Seas have to offer in terms of color, you will find out in Sect. 6.2 on the topic of "nighttime color spectacle on the beach". From my

---

Supplementary Information The electronic version of this chapter contains additional material, which can be accessed via the following link https://doi.org/10.1007/978-3-662-69575-3_5. The videos can be played by clicking on the DOI link in the legend of a corresponding figure, or by scanning this link with the SN More Media App.

141

experience, a UV flashlight with a wavelength of 365 nm (UV A) is best (example: Alonefire X901UV 365 nm UV flashlight, reference: Amazon, approx. 35 €). It is reliable, durable, and rechargeable via USB. I always have this UV lamp with me on all my trips.

5.1 With the UV Lamp in the Vegetable Garden and in the Kitchen

Although grass and many other green plants, when irradiated with white or blue light in the dark, glow red in a way that is not visible to our eye, the situation is quite different with UV light. In Sect. 2.1.5 I explain the red fluorescence of the green leaf pigment chlorophyll [1, 2]. A chlorophyll solution glows red when irradiated with white light. However, a green leaf, a meadow, or green vegetables irradiated with white light do not fluoresce red. But: Irradiation with *UV light* still produces a fluorescence glow, as graphically depicted in Fig. 5.1a.

UV light has much more energy than blue, green, or red light. The white light that we see, which is composed of the spectral colors, extends with wavelengths from about 750 nm to around 380 nm (nm = nanometers). UV light has a shorter wavelength than blue light and is divided into three categories (Fig. 5.1b). UV-A radiation ranges from 380 to 315 nm and UV-B light from 315 to 280 nm [3, 4]. 100 nm shorter wavelength than blue light may not sound like much, but you will notice how quickly UV-B radiation can cause sunburn when sunbathing. White or blue light, on the other hand, does not. Please always remember: The shorter the wavelength, the more

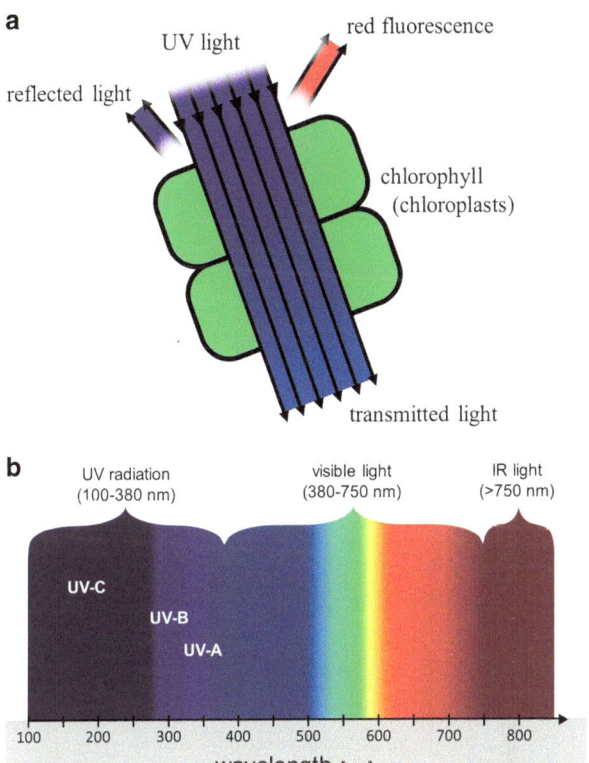

Fig. 5.1 (**a**) Path of UV radiation through a green plant or chlorophyll with the formation of a red fluorescence or a reflection. (**b**) Spectrum of light. Graphic: Melvin Müller

energy is contained in the light wave [5]. UV-C radiation (218–100 nm) is even more energy-rich, but fortunately it is completely absorbed by the atmosphere [3, 4]. If you now illuminate a plant with pure UV light, it receives an "excess" of light with a short wavelength. This "overdose" of light energy is quickly and easily rid by the irradiated plant by emitting light, i.e., fluorescence formation.

5.1.1 Experiment: The Pink Glow of Zucchini and Lamb's Lettuce

You will need

- Zucchini
- A handful of lamb's lettuce
- Knife
- UV flashlight (365 nm)

How it works

Illuminate a zucchini with white light. It appears dark green. Now darken the room or wait outside for darkness on the terrace or balcony and illuminate the zucchini with UV light. The green surface glows blood red (Fig. 5.2).

Cut a zucchini crosswise and illuminate the cross-section with UV light. A fantastically appearing pink ring shines (Fig. 5.3).

When lamb's lettuce is irradiated with UV light, its leaves glow in deep red (Fig. 5.4).

Fig. 5.2 A zucchini illuminated **a** with white light and **b** with UV light

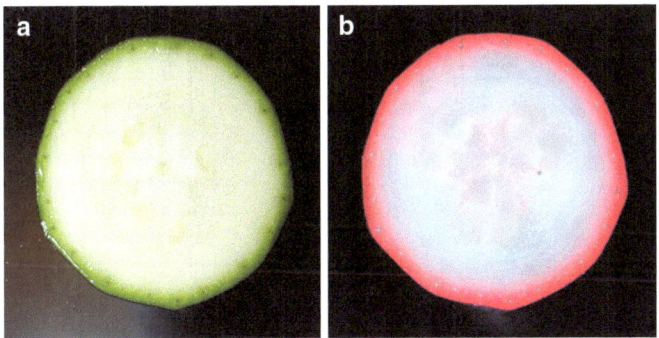

Fig. 5.3 A cut zucchini illuminated **a** with white light and **b** with UV light

Fig. 5.4 Lamb's lettuce illuminated **a** with white light and **b** with UV light

What's the science behind it?

The red glow is based on the red fluorescence of the green pigment chlorophyll (see Sect. 2.1.5). A large amount of chlorophyll is contained in the dark green lettuce leaves and in the zucchini skin. In addition, the leaves of the lamb's lettuce are relatively thin. White and blue light are too weak to stimulate the fluorescence visible to our eye. But UV-A light of wavelength 365 nm is strong and energy-rich enough to activate the chlorophyll molecules and thus "force" them to emit red light.

5.1.2 Experiment: Vegetable Smiley

You will need

- Zucchini
- Green pepper
- Some star moss
- Knife
- UV flashlight (365 nm)

How it works

Cut a slice from the zucchini and cut a smiley mouth out of the green pepper. Place mouth on slice. Two small star moss plants serve as eyes with lashes. The vegetable smiley is ready (Fig. 5.5). Your creativity knows no bounds!

Fig. 5.5 Vegetable smiley made of zucchini, pepper, and moss illuminated **a** with white light and **b** with UV light

5.1.3 Experiment: Green Bell Pepper— Unappealing Outside, Fascinating Inside

You will need

- Green bell pepper
- Knife
- UV flashlight (365 nm)

How it works

Illuminate a green bell pepper with white light. It appears dark green. Now darken the room or wait outside on the terrace or balcony for darkness and illuminate the pepper with the UV lamp. Unlike the zucchini, the green surface does not glow red, but only reflects the UV light (Fig. 5.6). Too bad! What's going on here? Why no red glow?

Now halve the pepper and illuminate the halves on the inside. Figure 5.7 shows the spectacular result. It almost looks like a cross-section of a heart.

What's the science behind it?

The texture of the outer shell layer (exocarp) of the pepper is quite hard and smooth like a wax layer. You also notice this when eating pepper slices, which is why some connoisseurs peel their bell peppers. The UV light cannot penetrate the shell and is only reflected. Therefore, the irradiated pepper in Fig. 5.6 appears blue-violet. There is no fluorescence glow.

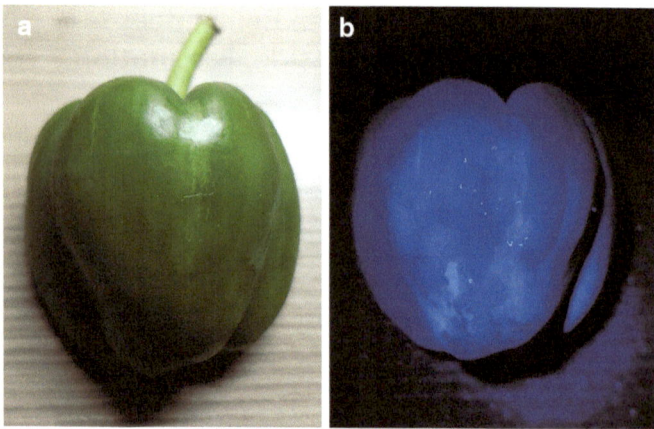

Fig. 5.6 A green bell pepper illuminated **a** with white light and **b** with UV light

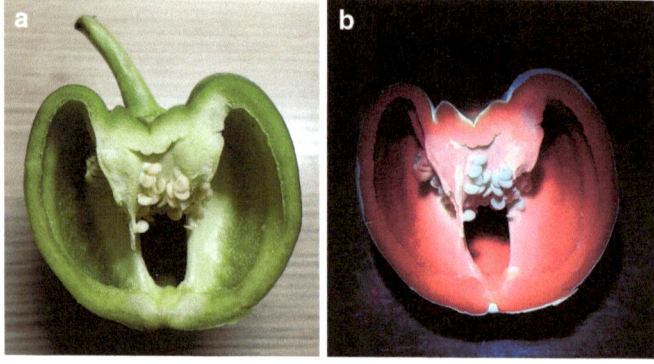

Fig. 5.7 A halved green bell pepper illuminated with **a** white light and **b** with UV light

However, this circumstance can also be used for a crea-tive-spooky fun for Halloween, and that without a huge, heavy pumpkin, but with a handy and tasty green bell pepper.

5.1.4 Experiment: Spooky Pepper

You will need

- Green pepper
- Knife or skin curette (Amazon, pharmacy)
- UV flashlight (365 nm)

How it works

Using a skin curette, carve a Halloween face into the surface of the pepper by scraping off the outer "shell" of the green pod down to the pulp. The sharp ring of a curette is ideally suited for this. Wipe the carved artwork briefly with a kitchen paper towel—done. Lights out, UV lamp on! The carved area glows blood red and looks really spooky, especially since the rest of the pepper appears in a pale blue (Fig. 5.8). Halloween can come. Afterwards, you can also eat the vitamin C-rich spooky pod as a healthy vegetable snack.

You can watch a "making-of video" of the spooky pepper by scanning the URL in Fig. 5.9 with your smartphone. It shows how I transform a green pepper into a "Frankenstein face" using a curette.

Try other types of vegetables and green plants as well! Not all green leaves fluoresce equally strongly. There are clear differences. While lettuce leaves, cucumbers, and zucchini glow intensely red, for example, tulip leaves, grasses, and conifers only show a weak red fluorescence. With a lit ivy, (young) cherry laurel, sage, or holly leaf, hardly any red fluorescence is visible to the naked eye. This

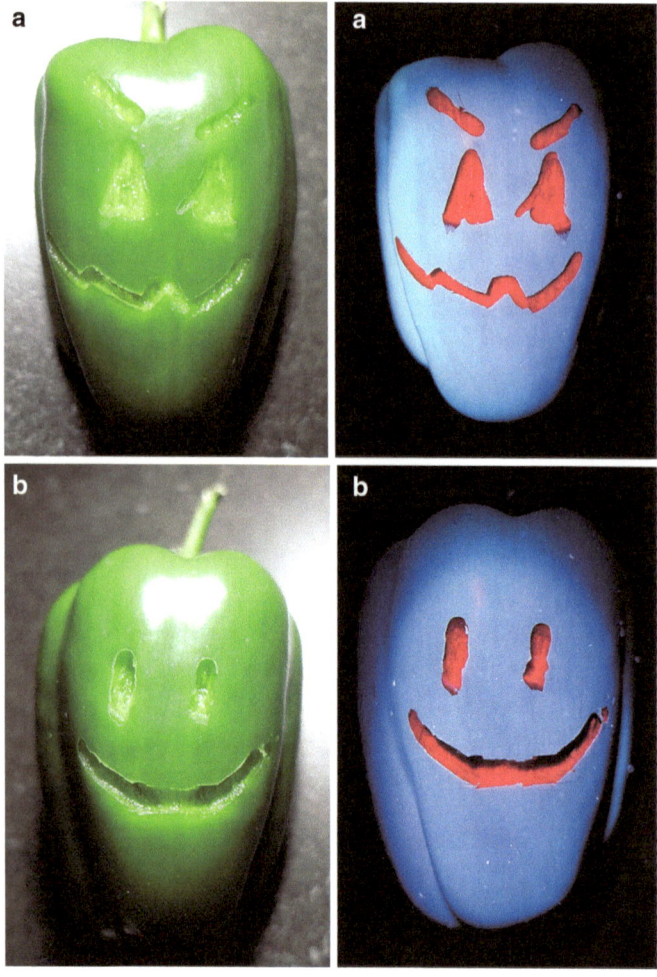

Fig. 5.8 **a** Spooky Halloween face and **b** Smiley carved into pepper and illuminated with white light (left) and with UV light (right)

is probably due to the texture and structure of the leaf surface. The red fluorescence of leaves under UV irradiation is also used to determine the "health status" of the respective plant (see Sect. 2.1.5).

Fig. 5.9 "Making-of" Spooky Pepper: The video shows the transformation of a green pepper into a "Frankenstein face". Music: Halloween by Hot Music (pixabay). Video Description: A face is carved into a green pepper with a curette, which fluoresces red in UV light URL: ▸https://doi.org/10.1007/000-d0n

5.1.5 Experiment: Radiant Kiwi

You will need

- Kiwi (green variant, not the golden one)
- Knife
- UV flashlight (365 nm)

How it works

Halve the kiwi and turn on the UV lamp in the dark. It looks a bit like an exploding supernova (Fig. 5.10).

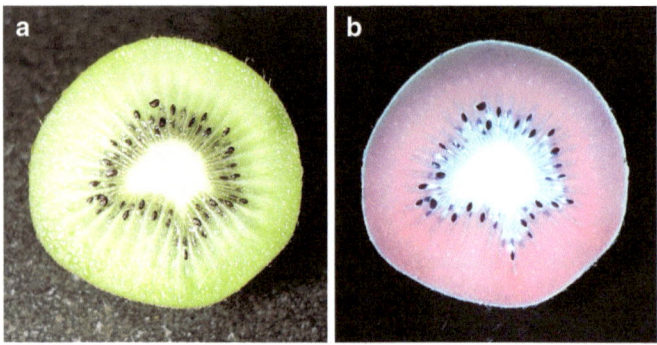

Fig. 5.10 Halved kiwi **a** illuminated with white light and **b** with UV light

5.1.6 Experiment: Sunset with Kiwi & Co.

You will need

- Kiwi (green variant, not the golden one)
- Some star moss
- A few pieces of yellow lichen
- Knife
- UV flashlight (365 nm)

How it works

Cut a slice from the middle of a kiwi and halve it. This serves as the sun setting on the sea horizon. Some star moss plants serve as little clouds and yellow lichen crusts are draped as sea waves. Lights out—sunset on! (Fig. 5.11).

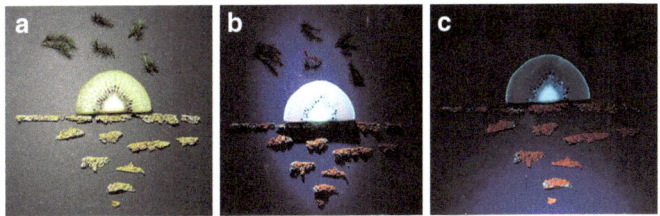

Fig. 5.11 Sunset scenario in the sea made of halved kiwi slice, star moss as little clouds and yellow lichens as waves **a** illuminated with white light and **b, c** with UV light

5.1.7 Experiment: Egg, why are you glowing so red?

You will need

- Brown eggs
- UV flashlight (365 nm)

How it works

If you illuminate brown eggs in the dark with UV light, a red fluorescence appears (Fig. 5.12).

What's the science behind it?

The brown eggshell contains, among other things, Protoporphyrin IX (Ooporphyrin), which fluoresces red when exposed to UV light. As with chlorophyll, a porphyrin molecule forms the basic structure [6, 7].

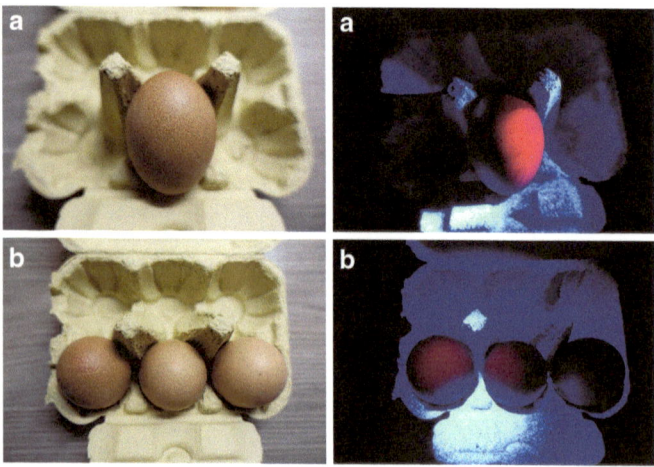

Fig. 5.12   a, b Brown eggs illuminated with white light (left) and with UV light (right)

5.2   With the UV Lamp in the Green, in the Forest, by the River

In the forest, there is so much to discover and marvel at with a UV lamp—you literally see nature in a different light [8, 9]. Mosses, mushrooms, stones, insects, but also garbage stand out brightly. Here I present some examples from my excursions. Either you remember the interesting places in daylight and illuminate them in the evening with UV light, or you go out with the UV lamp "armed" just to shine around randomly at dusk or in the dark. Usually, you discover all kinds of glowing things. But please don't forget to take a white LED lamp or your smartphone with you, so you know where you are going and above all see what is fluorescing here and there. From my experience, most luminous phenomena can already be recognized well in the twilight. It doesn't always have to be pitch dark outside.

5.2.1 Experiment: No moss, nothing (red) going on

You will need

- Moss plants in the forest, in the park
- UV flashlight (365 nm)
- LED lamp

How it works

If you discover beautiful moss areas in the forest during the day, remember the spots. Return to this location at dusk or in darkness and let the moss glow beautifully red. Figure 5.13a shows a star moss cushion in the Black Forest. With simultaneous illumination with white and UV light, you can nicely see the transition from green to red (Fig. 5.13b). In Fig. 5.14 you can see a silvery-blue shining tree stump that is overgrown with moss. I discovered it in the English Garden in Munich. It looks a bit like blood vessels around an organ.

Fig. 5.13 **a** Swan neck star moss (*Mnium hornum*) in the Black Forest illuminated with white light (left) and with UV light (right). **b** Simultaneous illumination of star moss with white and UV light. The transition from green to red is fluid

Fig. 5.14 Moss on a tree stump in the English Garden Munich illuminated **a** with white light and **b** with UV light

5.2.2 Experiment: Mushroom Beauty or Dr. Jekyll & Mr. Hyde

Here the motto is: Off to the mushrooms! Go (with your children) on a mushroom hunt with the UV lamp! But not necessarily to collect edible mushrooms ("boring"), but to discover glowing mushrooms ("cool"). Of course, you can also look for edible mushrooms. However, neither porcini mushrooms nor chestnuts fluoresce. Neither do champignons. One might suspect fluorescence in the bright red fly agarics, but here too: no show. However, non-fluorescence can also be very useful, as it can help to distinguish the edible from the poisonous variant in at least *one* type of mushroom.

You will need

- Autumn forest
- UV flashlight (365 nm)
- LED lamp

How it works

I spotted the following mushrooms in daylight and "successfully" illuminated them with UV light in the evening. Figure 5.15 shows the poisonous Sulphur Tuft *(Hypholoma fasciculare)*, which I discovered in the Southern Black Forest. This mushroom fluoresces in bright green under UV light.

There is also an edible variety of the Sulphur Tuft, namely the Grey Sulphur Tuft *(Hypholoma capnoides)*, which does *not* glow under UV irradiation. This is considered an excellent edible mushroom, even a delicacy [10, 11]. So we are dealing here with a kind of Dr. Jekyll and Mr. Hyde. Both Sulphur Tufts prefer to grow on dead wood or tree stumps, but the poisonous variety exclusively on coniferous wood. The risk of confusion is quite high for laymen, but a check with the UV lamp provides a clear result.

What's the science behind it?

The green-leaf sulfur head contains hypholomines and fasciculines, which fluoresce green in UV light [9]. Even seasoned chemists will have heard little of these exotic substances. Hypholomin A and B are yellowish pigments responsible for the green fluorescence [12]. Fasciculin A and B, apart from an additional two carbon and two hydrogen atoms (-HC=CH double bond), have the identical molecular structure as hypholomines A and B [13]. The variety of ingredients in mushrooms is enormous and most of them are still waiting for their discovery and structural elucidation. Which substances also glow in UV light remains to be deciphered. Literature provides an insight

Fig. 5.15   a–c Sulphur Tuft on dead wood illuminated with white light (left) and with UV light (right). Poisonous!

into the molecular richness and complexity of mushroom pigments [14].

The common sulfur polypore is also considered an excellent edible mushroom and prefers to grow on oak

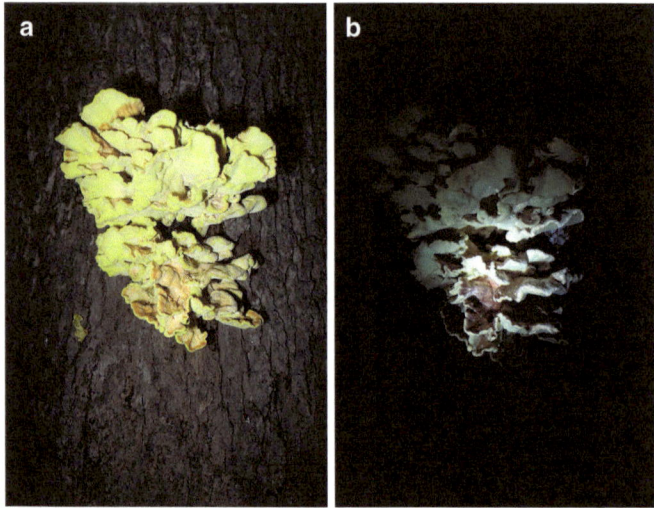

Fig. 5.16 Common sulfur polypore *(Laetiporus sulphureus)* on an oak in the Great Garden Dresden illuminated **a** with white light and **b** with UV light. Edible

trunks. In this case, the edible mushroom glows in UV light without any risk of confusion (Fig. 5.16).

Among wood producers and park gardeners, however, this tree fungus is very much feared and extremely unpopular, as it attacks the heartwood as an aggressive parasite and causes tree rot [15, 16]. It particularly likes to grow on old oaks, which it kills by internal crumbling and hollowing out. In the Great Garden in Dresden, I was able to observe this drama myself when I went back to photograph this mushroom on the old oak and found that the entire bark, including the mushroom, had been planed off where the mushroom was growing on the tree. At first, I was horrified, thinking of vandalism or animal bite, but I have since read that the sulfur polypore is a tasty but also a murderous villain. Another Dr. Jekyll-Mr. Hyde mushroom. The park gardeners have therefore struck to save the valuable oak or at least tried to prolong its life.

What's the science behind it?

The fluorescence-capable pigments are based on similar structures to those of the sulfur heads.

The turkey tail *(Trametes versicolor),* also a tree fungus, is considered a vital and medicinal mushroom against cancer, rheumatism and the like in Asia and in naturopathy, and is also supposed to be good for the immune system against viral infections (Fig. 5.17). It comes in all possible colors: black, white, yellow, brown, with and without ring patterns. Here, it is used in dried form as a decorative mushroom in floristry for arrangements and advent wreaths. Its consistency is like an old shoe sole and in the dried state it becomes rock hard [17, 18].

The most beautiful and breathtaking glow, however, is shown by the pink radish bonnet *(Mycena rosea)* [19, 20]. As soon as the UV light illuminates the hat in the dark, the mushroom lights up azure blue, as if it were lit from the inside (Fig. 5.18). Like a switched-on lamp with a blue shimmering shade. Wonderful!

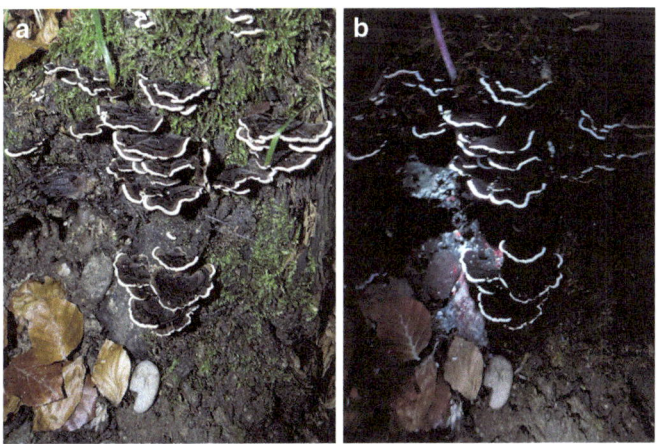

Fig. 5.17 Turkey tail *(Trametes versicolor)* in the Black Forest illuminated **a** with white light and **b** with UV light. Inedible

Fig. 5.18 Pink radish bonnet *(Mycena rosea)* illuminated **a** with white light and **b** with UV light (right)

What's the science behind it?

Which ingredients trigger the beautiful blue fluorescence is not yet known. It could be russopteridin and russulumazin, two pteridin derivatives. In other types of mushrooms, this class of substances is responsible for blue and yellow-green fluorescence effects [9].

5.2.3   Experiment: At the Stream, River, Pond

You will need

- Stream or river
- UV flashlight (365 nm)
- LED lamp

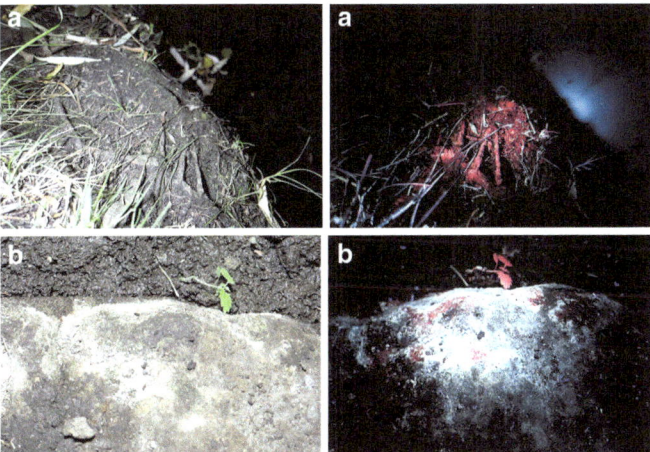

Fig. 5.19 **a** Riverbank illuminated with white light (left) and with UV light (right). **b** Stone and plant in the riverbed illuminated with white light (left) and with UV light (right)

How it works

Walk along the shore and just start shining! The inconspicuous suddenly becomes visible. Figure 5.19a shows a typical riverbank. The blue area in the right picture is the Ilmenau River in Lüneburg. In Fig. 5.19b you can see a stone and a young plant full of chlorophyll in the riverbed. Chlorophyll can also be seen on the stone due to the red fluorescence, originating from mosses or algae.

5.2.4 Experiment: Red Duckweed

You will need

- Standing body of water with duckweed

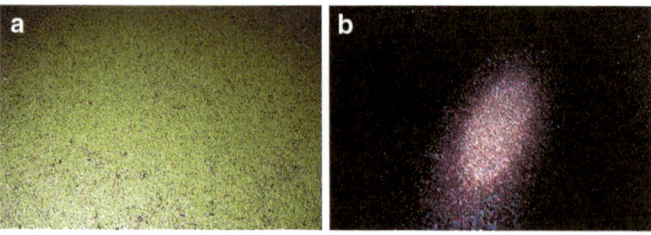

Fig. 5.20 Duckweed (*Lemna*) in a pond **a** illuminated with white light and **b** with UV light

- UV flashlight (365 nm)
- LED lamp

How it works

When illuminating duckweed (Fig. 5.20), you need to get relatively close to the water to achieve a good effect. So be careful! Don't fall into the water in the dark.

5.2.5 Experiment: Detecting Trash

Here I show you two examples of trash in the forest, where it absolutely does not belong. With the UV lamp, I saw the bright fluorescence from afar. At first, I thought I had discovered something special, but no such luck.

You will need

- Forest, park
- UV flashlight (365 nm)
- LED lamp

How it works

In the dark, shine your UV lamp off the paths into the bushes and undergrowth. You wouldn't believe how much trash and waste you discover. Guess what object is shown in Fig. 5.21a! In Fig. 5.21b it is obvious.

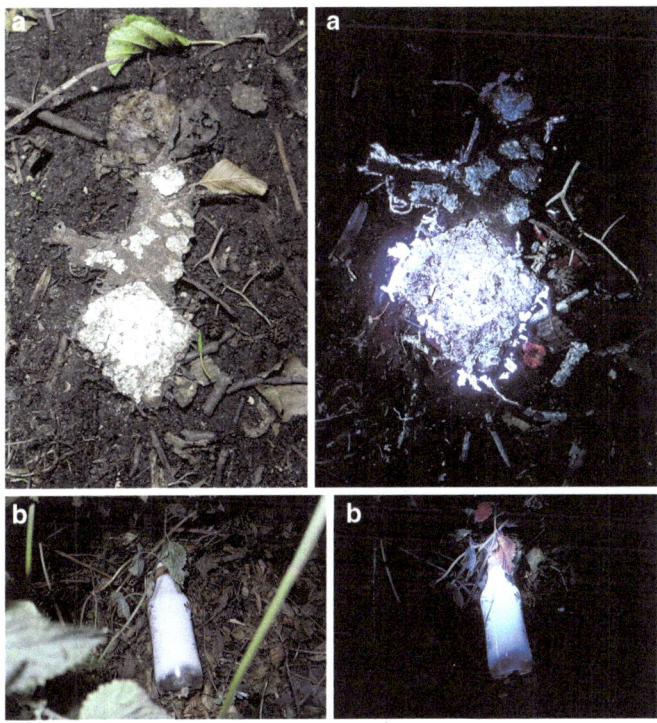

Fig. 5.21 **a** A rotten sneaker illuminated with white light (left) and with UV light (right). **b** A bottle of "Malibu Original Coconut Liqueur" in the bushes illuminated with white light (left) and with UV light (right)

5.3   Luminous Lichens

In Sect. 3.4.1 I have already described the Common Orange Lichen *(Xanthoria parietina)* [21], whose pigment parietin fluoresces intensely orange under UV light (Fig. 5.22). Orange lichens are often found on trees with sufficient nutrient supply, for example near fields and farmlands, but also often on street trees, where dogs do their small and large "business". Good fertilizer for the tree, good conditions for the lichen. Due to the good and SO_2-reduced air nowadays, it also thrives on city trees and even on stones or garden gates.

The pigments of the sulfur lichens also glow under UV irradiation. Sulfur lichens belong to the crustose lichens, are bright yellow and have a fine-grained, water-repellent (hydrophobic) surface. They only thrive in a very pure air environment on large silicate rocks in moist valleys, such as the Elbe Sandstone Mountains. There are mainly two types of sulfur lichens [22]: The Rock Sulfur Lichen *(Chrysothrix chlorina)* and the Yellow-Fruited Sulfur Lichen *(Psilolechia lucida)*..

Fig. 5.22 Common Orange Lichen on a linden trunk, illuminated with white light (left) and with UV light (right). Length of the orange lichen: 9 cm

5.3.1 Experiment: Yellow Sulfur Lichens— Orange Luminescence

You will need

- Suitable location (e.g. Elbe Sandstone Mountains)
- UV flashlight (365 nm)
- LED lamp

How it works

Illuminate a sulfur lichen in the dark with UV light and green-yellow or yellow becomes orange (Fig. 5.23).

What's the science behind it?

The Yellow-Fruited Sulfur Lichen is rather green-yellow and contains the dye Rhizocarpic acid, while the Rock Sulfur Lichen stands out brightly yellow due to its yellow pigment Vulpinic acid. Both Rhizocarpic acid and Vulpinic acid belong to the group of Pulvinic acid dyes (Lactone pigments), which fluoresce in the range of 550–570 nm. Both yellow, color-determining pigments therefore fluoresce dark orange under UV light. Figure 5.23a shows the Yellow-Fruited Sulfur Lichen on a rock of the Bastei in the Elbe Sandstone Mountains. The chlorophyll in the lush green moss fluoresces blood red, while the green-yellow pigment Rhizocarpic acid of the sulfur lichen glows orange. In Fig. 5.23b, the yellow Vulpinic acid of the Rock Sulfur Lichen fluoresces in dark orange. The water droplets on the hydrophobic surface can be seen much better in both images under UV light due to

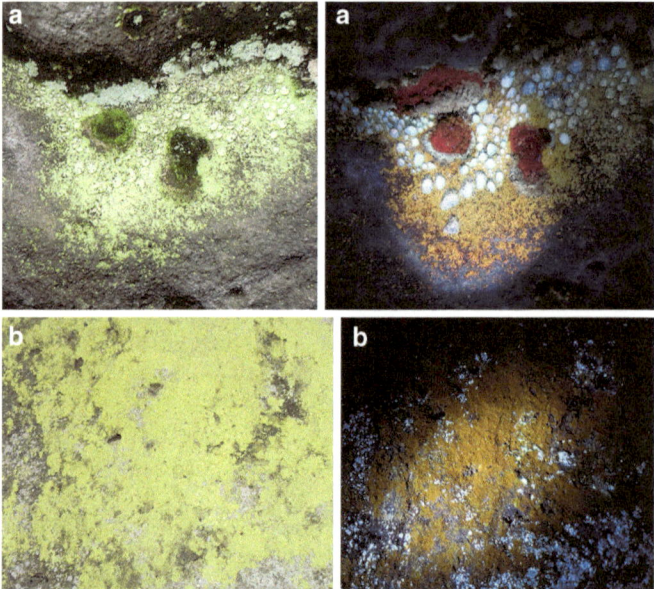

Fig. 5.23 **a** Rain-wet Yellow-Fruited Sulfur Lichen on a rock crevice in the Elbe Sandstone Mountains, illuminated with white light (left) and with UV light (right). **b** Rain-wet Rock Sulfur Lichen on a rock in the Kirnitzsch Valley in the Elbe Sandstone Mountains, illuminated with white light (left) and with UV light (right)

the bluish reflection than in white light. Due to the lotus effect, the water droplets have formed into beads (see Sect. 1.2).

5.4   Small, Glowing Ghosts

The woodlouse *(Porcellio scaber)* lives with its 12–18 mm tininess in damp dark places, among other things under stones or dead wood, under flower pots or objects in the garden, in the gravel bed around the house wall, in the raised bed, in the compost heap [23]. Woodlice have seven pairs of legs, a seven-segmented chitin shell and consume dead organic substances. Each of you has seen them somewhere before. Disgusting? No, they are very useful little creatures—and they glow under UV light.

5.4.1 Experiment: Turning Woodlice into Ghosts

You will need

- Woodlice
- UV flashlight (365 nm)
- LED lamp

How it works

If you have discovered woodlice or know where they are hiding, then go hunting in the dark with the UV lamp in the evening. Please only briefly illuminate these little creatures to spare them unnecessary light stress. The woodlice glow in a ghostly appearing blue-white light. Suitable for Halloween. I observed at our terraced house that woodlice massively flee from the gravel bed at the house wall upwards to escape the floods after a heavy rain shower. The house wall was covered with woodlouse refugees. There I could illuminate and photograph them in peace (Fig. 5.24). In the close-up of Fig. 5.24b and c, you can even see the seven shell segments on the right woodlouse.

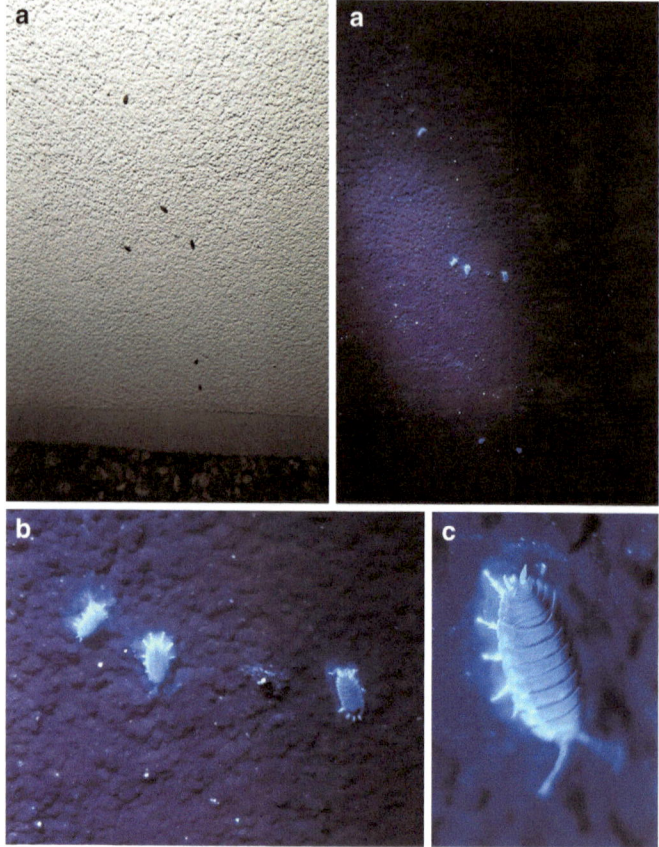

Fig. 5.24 **a** Woodlice on the house wall illuminated with white light (left) and with UV light (right), **b** and **c** Close-ups

What's the science behind it?

The chitin of the "outer skin" (exoskeleton) has fluorescent properties and emits light with a wavelength of about 440 nm—in other words: blue light—[24, 25]. The blue glow effect is based on the presence of chitin, which is made up of thousands of acetylglucosamine units. It is similar to cellulose, but is much harder and more stable.

With the help of UV radiators, scorpions can also be tracked down to avoid nasty surprises during excursions or hikes in the habitats of these eight-legged creatures. Some spiders also fluoresce in blue [9].

5.5 Colorful Luminous Stones, Colorful Stone Lighting

The variety of minerals is immense and extremely complex [26]. As a layman, I know amethyst, agate, malachite & Co., but then it soon ends. Some specimens show fluorescence lighting, which will be briefly discussed here.

5.5.1 Experiment: Fluorescent Minerals

You will need

- Various minerals such as hyalite, calcite, wurtzite or fluorite
- UV flashlight (365 nm)

How it works
Admittedly: The minerals depicted and described here I did not find myself, but bought in the breathtaking mineral collection "terra mineralia" in Freiberg/Saxony in the local shop [27]. With a bit of luck, patience, knowledge and on the right mountain, fluorescent stones can still be found. Under UV light, green, red, pink, orange-yellow

Fig. 5.25 **a** Hyalite on feldspar (Namibia), **b** Calcite (Austria), **c** Wurtzite (Poland), **d** Fluorite on siderite (Ore Mountains) irradiated with white light (left) and with UV light (right)

and blue-violet fluorescence phenomena become visible. Figure 5.25 shows some examples.

What's the science behind it?

Most often, the fluorescence in minerals is caused by the incorporation of foreign metal ions, such as yttrium, cerium, manganese, europium, and uranium ions [28, 29].

Fig. 5.25 (continued)

The most common fluorescent mineral in Germany is fluorite (CaF_2), a compound of calcium and fluorine, also known as fluorspar. The siderite consists of iron carbonate ($FeCO_3$) and serves here as the basis of the fluorite crystals (Fig. 5.25d). Fluorite is mined in the Ore Mountains, in Vogtland and in the Upper Palatinate and can also be found there. Calcite is calcium carbonate ($CaCO_3$), very common and generally known in the form of lime or chalk (Fig. 5.25b). As a very complex silicate mineral, feldspar contains various metals in addition to silicon-oxygen units, such as sodium, potassium, calcium, barium, and aluminum. The hyalite on it is a variant of opal and consists of silicon dioxide (SiO_2), which often contains uranium-oxygen (uranyl, UO_2^{2+}) inclusions (Fig. 5.25a). Wurtzite (ZnS) mainly contains zinc and sulfur (Fig. 5.25c). Comprehensive knowledge about minerals and rocks can be found in the literature [26]. The website of the Sterling Hill Mining Museum, which further explains the fluorescence of various minerals, is also worth reading [30]. By the way, entire rocks fluoresce in the most wonderful colors both inside and outside the mine in New Jersey (USA).

5.6   UV + LED = Light

Driven by the climate and energy crisis, rapidly rising electricity prices, and resource conservation, each and every one of us has a multitude of LED light sources in use in our homes. Good so. Right so. Street and building lighting is also being and will increasingly be converted to LED [31]. With the help of a UV lamp, one can trace the chemical inner workings of an LED. This now has nothing to do with nature in the green, but in my opinion, it is a beautiful and interesting experiment.

5.6.1   Experiment: Electric Trick with LED Lamps

You will need

- Various LED light sources e.g. filament LED, G9 Bi-Pin LED
- UV flashlight (365 nm)

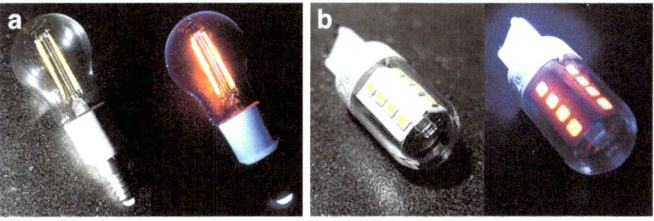

Fig. 5.26   **a** LED filament lamp, **b** G9 Bi-Pin LED each irradiated with white light (left) and with UV light (right)

How it works

For this simple experiment, you don't even have to unscrew or pull out the LED light sources, but can directly illuminate them with your UV lamp. To do this, please turn off the light and darken the room. Figure 5.26 shows how the individual LED points or filaments light up brightly in UV light.

What's the science behind it?

The "heart" of every LED light source is indium gallium nitride (InGaN), a semiconductor made of the elements indium, gallium, and nitrogen [32]. Depending on the composition, the InGaN emits ultraviolet or violet-blue light when electricity is supplied. To convert these wavelengths, which are not visible to our eyes, into white, warm, "visible" light, an additional "phosphor" is required. Almost all commercially available LED light sources are based on the COB (= Chip-on-Board) design. The LED sits directly on a mini circuit board and is then covered with a plastic mass that contains the phosphor [32, 33].

This is the intensely yellow-orange phosphor YAG, a complex compound of yttrium, aluminum, and oxygen with the formula $Y_3Al_5O_{12}$, to which cerium^{3+}- or other metal ions have been added [32]. YAG is the abbreviation for yttrium aluminum garnet, where "garnet" refers to a specific group of minerals [32, 34]. The UV light is absorbed by the YAG phosphor and loses energy in the process, so that the emitted fluorescence is shifted to lower-energy, i.e., longer-wavelength light—into the rainbow colors, i.e., into white light. In every LED, you can see the yellow dots or threads of the YAG:Ce^{3+} phosphor [32, 33].

Background

Filament LED

The first filament LEDs come from Japan and were invented in 2008. Typical filament or filament LEDs in retro design have been on the market since 2015 and are very popular. Each filament consists of a thin strip of glass, on which about 30 tiny LED chips in COB design are mounted side by side and connected in series. The approximately 4 cm long LED filament is finally coated with the yellow fluorescent material YAG:Ce^{3+}. The blue-violet light emitted by the LEDs is thus converted into warm yellow light.

References

1. A. Korn-Müller, *Warum Gras nicht rot leuchtet*, Nachr. Chem. 70, **2022**, pp. 18–21.
2. L. Urry, M. Cain, S. Wasserman, P. Minorsky und J. Reece, M., *Campbell Biologie*, 11., aktualisierte Aufl., Pearson Verlag Deutschland, München, **2019**, pp. 260–261.
3. C.-D. Schönwiese, Klimatologie, 5., überarbeitete und aktualisierte Aufl., Eugen Ulmer, Stuttgart, **2020**, pp. 29 und 276.

4. D. Weiß und H. Brandl, *Fluoreszenzfarbstoffe in der Natur, Teil 1*, Chem. Unserer Zeit, 47, **2013**, p. 52.

5. A. Korn-Müller, *Mit Zollstöcken und Springseil zum Regenbogen*, Nachr. Chem. 69, **2021**, pp. 28–31.

6. J. M. Berg, J. L. Tymoczko, G. J. Gatto jr. und L. Stryer, *Stryer Biochemie,* 8th edn., Springer Spektrum, Heidelberg, **2018**, pp. 866–867.

7. D. Weiß, E. Täuscher und H. Brandl, *Die bunte Welt der Porphyrine*, Chem. Unserer Zeit, 53, **2019**, pp. 12–21.

8. D. Weiß und H. Brandl, *Fluoreszenzfarbstoffe in der Natur, Teil 1*, Chem. Unserer Zeit, 47, **2013**, pp. 50–54.

9. a) D. Weiß und H. Brandl, *Fluoreszenzfarbstoffe in der Natur, Teil 2*, Chem. Unserer Zeit, 47, **2013**, pp. 122–131. b) http://www.chemie.uni-jena.de/institute/oc/weiss/start.html (last acccss: Oct 18, 2024).

10. https://fundkorb.de/pilze/hypholoma-capnoides-grau-bl%C3%A4ttriger-schwefelkopf (last access: Oct 18, 2024).

11. http://tintling.com/pilzbuch/arten/h/Hypholoma_capnoides.html (last access: Oct 18, 2024).

12. https://roempp.thieme.de/lexicon/RD-08-02625?linkSource=TIB (last access: Oct 18, 2024).

13. https://roempp.thieme.de/lexicon/RD-06-00255?searchterm=fasciculine&context=search (last access: Oct 18, 2024).

14. J. Velisek and K. Cejpek, *Pigments of Higher Fungi: A Review*, Czech J. Food Sci., 29, **2011**, pp. 87–102.

15. http://tintling.com/pilzbuch/arten/l/Laetiporus_sulphureus.html (last access: Oct 18, 2024).

16. https://fundkorb.de/pilze/laetiporus-sulphureus-schwefel-porling (last access: Oct 18, 2024).

17. https://fundkorb.de/pilze/trametes-versicolor-schmetterling-stramete (last access: Oct 18, 2024).

18. http://tintling.com/pilzbuch/arten/t/Trametes_versicolor.html (last access: Oct 18, 2024).

19. https://fundkorb.de/pilze/mycena-rosea-rosa-rettich-helmling (last access: Oct 18, 2024).

20. http://tintling.com/pilzbuch/arten/m/Mycena_rosea.html (last access: Oct 18, 2024).

21. V. Wirth und U. Kirschbaum, Flechten einfach bestimmen, 2., aktualisierte Aufl., Quelle & Meyer Verlag, Wiebelsheim, **2017**, p. 35.
22. V. Wirth und U. Kirschbaum, Flechten einfach bestimmen, 2., aktualisierte Aufl., Quelle & Meyer Verlag, Wiebelsheim, **2017**, p. 244.
23. https://bodentierhochvier.de/steckbrief/porcellio-scaber/ (last access: Oct 18, 2024).
24. M. D. Rabasovic, D. V. Pantelic, B. M. Jelenkovic et al., *Nonlinear microscopy of chitin and chitinous structures: a case study of two cave-dwelling insects*, J. Biomed. Opt., 20, **2015**, pp. 016010–1–016010–10.
25. Q. Dong, W. Qiu, L. Li, N. Tao et al., *Extraction of chitin from white shrimp (Penaeus vannamei) shells using binary ionic liquid mixtures*, J. Ind. Eng. Chem.,120, **2023**, pp. 529–541.
26. G. Markl, *Minerale und Gesteine*, 3. ed., Springer Spektrum Verlag, Berlin Heidelberg, **2014**.
27. https://terra-mineralia.de (last access: Oct 18, 2024).
28. https://www.mineralienatlas.de/lexikon/index.php/ MineralData?lang=de&mineral=Hyalit (last acess: Oct 18, 2024).
29. https://www.mineralienatlas.de/lexikon/index.php/ Fluoreszenz (last access: Oct 18, 2024).
30. https://www.sterlinghillminingmuseum.org/fluorescence (last access: Oct 18, 2024).
31. R. Heinz, *Grundlagen der Lichterzeugung*, 5., erweiterte Aufl., **2014**, Highlight Verlag, Rüthen, p. 105.
32. F. Baur und T. Jüstel, *Weiße Leuchtdioden als moderne Leuchtmittel: Anorganische Materialien*, Chem. Unserer Zeit, 56, **2022**, pp. 220–231.
33. R. Heinz, *Grundlagen der Lichterzeugung*, 5., erweiterte Aufl., **2014**, Highlight Verlag, Rüthen, pp. 92–108.
34. Holleman/Wiberg, *Anorganische Chemie*, vol. 2, 103rd edn., Walter de Gruyter Verlag, Berlin, **2017**, p. 1787.

6

On the Beaches of the North and Baltic Seas

Abstract The North and Baltic Seas have great ecological and touristic significance. With their resorts and islands, they attract hundreds of thousands of vacationers. In addition to relaxation, stimulating climate, good air, beaches and dunes, the coasts offer all sorts of natural experiences—mostly during the day. But even at night, one can experience numerous colorful phenomena. From the breathtaking, blue bioluminescence of the sea to the color spectacle in UV light: red, yellow, blue, and orange glowing algae, blue fluorescent crab shells and jellyfish. In daylight, your children might enjoy a handmade, small sand avalanche or you might enjoy finding chalk fossils, pieces of amber, and lucky stones.

Supplementary Information The electronic version of this chapter contains additional material, which can be accessed via the following link https://doi.org/10.1007/978-3-662-69575-3_6. The videos can be played by clicking on the DOI link in the legend of a corresponding figure, or by scanning this link with the SN More Media App.

6.1   Marine Luminescence—Cold Fireworks in Blue

Chemiluminescence ("cold light") always occurs when the conversion of chemical energy into light takes place during a chemical reaction. This is also referred to as chemically excited fluorescence [1]. If it is an enzymatically catalyzed chemiluminescence based on the luciferin-luciferase system in living organisms, it is referred to as bioluminescence. Prime examples are the local fireflies *(Lamprohiza splendidula),* the great fire beetle *(Lampyris noctiluca),* the North American fire beetle *(Photinus pyralis),* the marine luminescent microorganisms *(Noctiluca miliaris/scintillans)* as well as various deep-sea marine animals such as jellyfish, squids, fish, and shrimps [1, 2].

With a bit of luck, the blue-glowing *Noctilucae* can also be observed in the North Sea, for example in Germany and Belgium. These microscopic marine algae belong to the dinoflagellates, are single-celled, between 200–2000 μm small and can be mechanically stimulated to bioluminescence [3]. For this, it is best to walk through the water-filled sandbanks at low tide at night. Figure 6.1 shows such a typical sandbank.

If you are looking for marine luminescent creatures, then pay attention to sandbanks as shown in Fig. 6.2.

Fig. 6.1   Sandbank at low tide on the Belgian North Sea coast

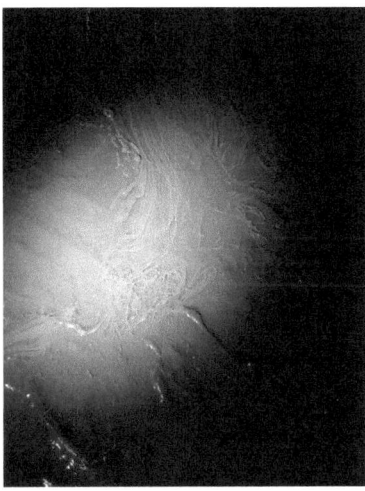

Fig. 6.2 Sandbank at low tide on the Belgian North Sea coast at night, illuminated with an LED lamp. The pink streaks are teeming with marine luminescent creatures

What appears so inconspicuous in mouse gray, rather disgustingly muddy, pink-brownish and unattractive floating on the water, turns out to be a breathtaking color spectacle. Since the bioluminescence is relatively weak, it must be pitch dark outside.

6.1.1 Experiment: Luminous Footprints in Blue (North Sea)

You will need

- North Sea beach
- Some courage to trudge through the sandbanks in pitch darkness

How it works

Just step into the sandbank mud, as shown in Fig. 6.2, and you will experience a marvel! Stimulated by mechanical movements, such as stomping of the foot, wave movements by hand or foot, and by water splashes, a spectacular blue firework is ignited. In fractions of a second, the wave movements are transmitted and lead to an "explosive" illumination in magnificent blue (Fig. 6.3).

It almost feels like the miraculous planet "Pandora" from the movie "Avatar", where every footprint makes the ground glow. Don't worry, these organisms are so tiny that you don't kill them when you step on them. You can

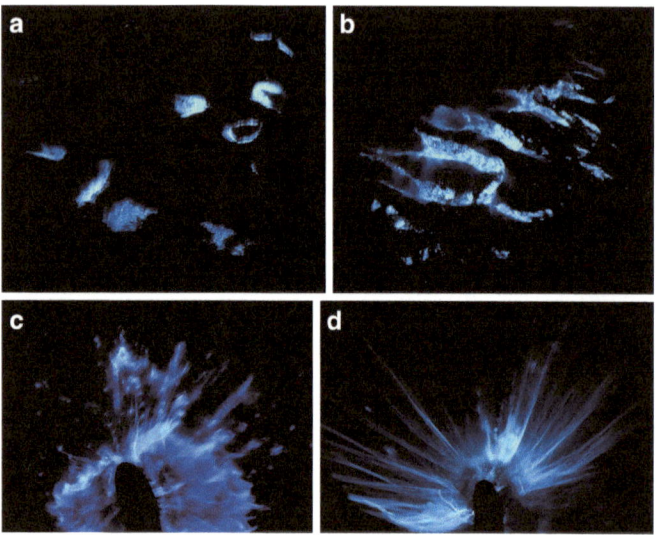

Fig. 6.3 Blue bioluminescence of the marine luminescent dino-flagellates *(Noctiluca scintillans)* through mechanical stimulation on a sandbank of the Belgian North Sea coast in the month of April. **a, b** Bioluminescence transmitted in the wave furrows of the sandbank in a fraction of a second. **c, d** Light reaction triggered by the foot at the edge water of a sandbank

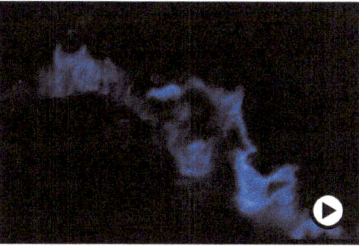

Fig. 6.4 The seawater glows blue, Part 1! The video shows the blue bioluminescence of the marine luminescent dinoflagellates on the North Sea coast. Video Description: By stomping the foot in a sandbank at night, the marine luminescent dinoflagellates are stimulated and glow blue URL: ▸ https://doi.org/10.1007/000-d0p

even repeat the stimulation to the blue, cold light as often as you like. This is also necessary, because photographing this beautiful effect is not easy. In this section, I have stored two of my most beautiful marine glow videos for you. Simply scan the URLs in Figs. 6.4 and 6.5 with your smartphone and enjoy the spectacular blue, cold light. You can also hear the sea rushing and the wind whistling. North Sea at it's best!

What's the science behind it?

The marine luminescent microorganisms (*Noctiluca scintillans*) are dinoflagellates, belong to the animal plankton (zooplankton), and are eukaryotic single cells with a size of 200–2000 μm. They produce bioluminescence, which is triggered by mechanical stimulation, such as wave movements, wind, or other living beings. The thereby "pressed in" cell membrane leads to a chemical, enzymatic reaction, in which blue light with a wavelength of about 475 nm is emitted.

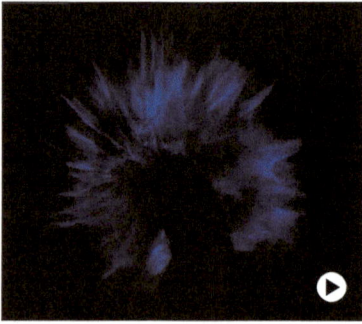

Fig. 6.5 The seawater glows blue, Part 2! The video shows the blue bioluminescence of the marine luminescent dinoflagellates on the North Sea coast. Music: Aquamarin—The Shimmer of Blue Ocean Ambient chill music by Julius H (pixabay) URL: ▸ https://doi.org/10.1007/000-a6d

Most bioluminescence reactions in the animal kingdom involve an enzyme-catalyzed conversion of luciferin "luminous substances" using the enzyme luciferase [1, 2]. In this process, the luciferin, which can have very different molecular structures, is put into an excited state via oxygen uptake and carbon dioxide release, which finally transitions back to the ground state by emitting light. The bioluminescence of the marine luminescent dinoflagellates *(Noctiluca scintillans)* is based on a similar reaction, the details of which, however, have not yet been clarified [2]. It is certain, however, that it is triggered by mechanical (and chemical) stimuli. The flash-like lighting serves both to deter predators and as a "warning color" [2].

The fantastic marine luminescent plankton can also be observed in the Baltic Sea, but only in a short time window. With a little luck, they can be found in high summer, around mid-August. The "Nordkurier" of the Neubrandenburg newspaper shows spectacular photos of blue glowing waves in Lubmin and on the Greifswalder Bodden on its online page, taken in August 2020. The

"Ostsee-Zeitung" from Mecklenburg-Vorpommern also reports online in August 2022 about marine luminescence in Lubmin. Unfortunately, I have not yet been able to admire marine luminescence in the Baltic Sea. Too bad. Recently, even night-time excursions for tourists have been offered, for example in Neuharlingersiel: "Marine luminescence—strolling in the nocturnal sea of lights" is what it's called [4]. The National Park House Wurster North Sea Coast near Bremerhaven also offers night excursions every year in summer [5].

The occurrence of marine luminescence on the North and Baltic Sea coasts unfortunately always remains a game of chance, as wind and weather, water temperature, and ocean currents play a significant role. However, conservationists view a too strong spread of the sea sparkle creatures due to over-fertilization of the seas very critically, as it could negatively change the food chains. The massive reproduction of these single cells is also referred to as "red tide" and often causes a mass die-off of other creatures due to excessive oxygen consumption and excretion of toxic substances [2].

6.2   Nighttime Color Spectacle on the Beach—The Dark Side of the Sea

In this section, I would like to introduce you to the "dark" side of the beaches, because at night you can experience numerous colorful phenomena. If you are going on vacation to the North or Baltic Sea (or another sea), I can only highly recommend taking a UV lamp with you. From my experience, a UV flashlight with a wavelength of 365 nm (UV-A) works best (example: Alonefire X901UV 365 nm UV flashlight, reference: Amazon, approx. 35 €). It is reliable, durable, and rechargeable via USB. I always have the UV lamp with me on all my trips. A night walk on the beach is worth it and your kids will experience something very special—despite the fear of darkness, which sometimes needs to be overcome! What looks like old seaweed turns out to be a colorful spectacle and inconspicuous gray mud mutates into a colorful work of art of nature. Invisible structures glow in bright colors and shells, crab shells, and jellyfish show impressive fluorescence phenomena. For biology and chemistry teachers, a scientific excursion or "science week" at the North or Baltic Sea with a UV night hike would be a real and new experience. I was delighted to read by chance that the Domgymnasium Magdeburg has been conducting study and experimental weeks at the Wadden Sea for years with the bio and chemistry graduation level [6]—but still without UV light actions at night!

With my UV flashlight, I have been both on the Belgian North Sea (Oostduinkerke) and on the German Baltic Sea (Fischland-Darß, Greifswalder Bodden). Here are my most beautiful discoveries in the dark.

6.2.1 Experiment: Bright Red Algae Sludge (North Sea)

You will need

- North Sea beach
- UV flashlight (365 nm)

- LED lamp
- A bit of courage to trudge through the sandbanks in pitch darkness

How it works

Shine your UV lamp at low tide into the sandbanks. Where there is brownish-green algae sludge, a spectacular red fluorescence occurs, as can be seen in Fig. 6.6. But an orange fluorescence can also be admired in Fig. 6.6c.

What's the science behind it?

The tiny marine algae that have accumulated in the sludge of the sandbank in the wave furrows contain chlorophyll, which fluoresces intensely red in UV light. The piece of seaweed in Fig. 6.6c, which glows brightly orange, is probably sugar kelp, a brown alga *(Saccharina latissima)* [7]. The bladderwrack shown in Fig. 6.6d also belongs to the brown algae. Brown algae contain, in addition to chlorophyll, the dark red to dark brown pigment fucoxanthin, a carotenoid of the xanthophyll group [8]. It is a light-collecting molecule, absorbs light in the green wavelength range, and thus makes photosynthesis underwater, where naturally less sunlight reaches, more efficient because green light can penetrate deeper into the water than red light. In UV light, fucoxanthin fluoresces with a wavelength of about 630 nm in bright orange-red [8].

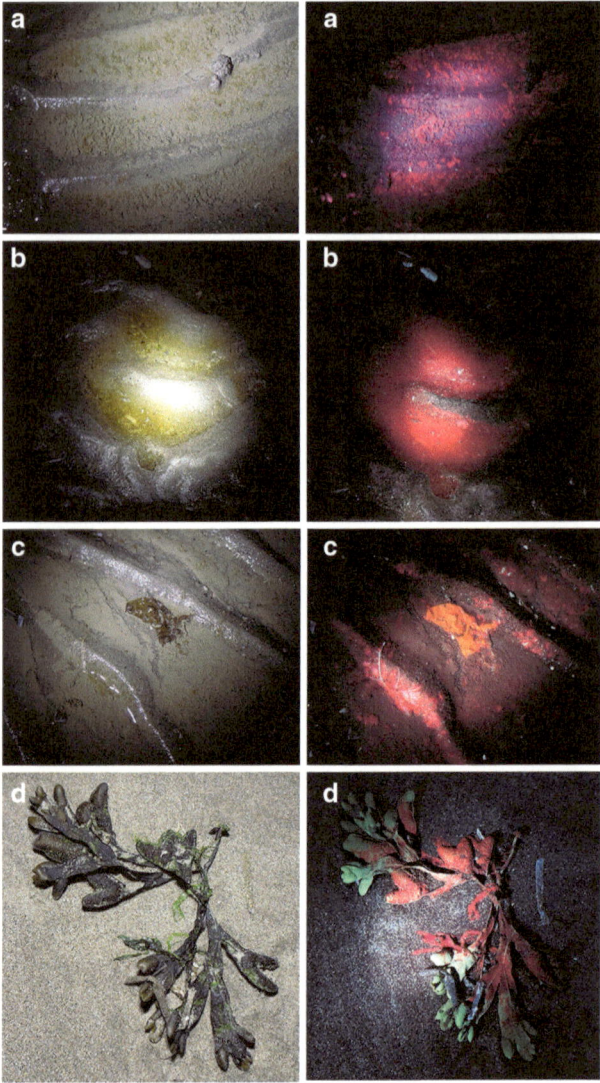

Fig. 6.6 Algae in the wave furrows of a sandbank, illuminated with white light (left) and with UV light (right). **a, b** Marine algae fluoresce in bright red. **c** In addition to the marine algae in red, a piece of seaweed glows brightly in orange. **d** Bladderwrack fluoresces in green and orange-red

6.2.2 Experiment: Sky Blue Crab Shell (North Sea)

You will need

- North Sea beach
- UV flashlight (365 nm)
- LED lamp
- Some courage to search the edge of the sandbanks in pitch darkness

How it works

Search the wash zone of the sandbanks at low tide with an LED light for discarded, washed-up carapaces of crabs. Some of them glow beautifully blue in UV light. Figure 6.7 shows a (dead) swimming crab and the discarded carapace of a swimming crab that I discovered on my evening walks. By chance, a quick North Sea shrimp also darted in front of my camera. In UV light, you can see it well.

What's the science behind it?

In Fig. 6.7a and b, it is a swimming crab *(Portunus holsatus),* in Fig. 6.7c it is a North Sea shrimp (Crangon crangon) [9]. The shells, legs, and claws of these crustaceans (Crustacea) consist mainly of the nitrogenous sugar chain molecule chitin (20–30%), the structural protein arthropodin or sclerotin (30–40%), and lime ($CaCO_3$, 30–50%) [10, 11]. From this "cement mixture," fibers are formed that layer on and within each other to form super-hard plates (cuticle). The blue glow effect is based

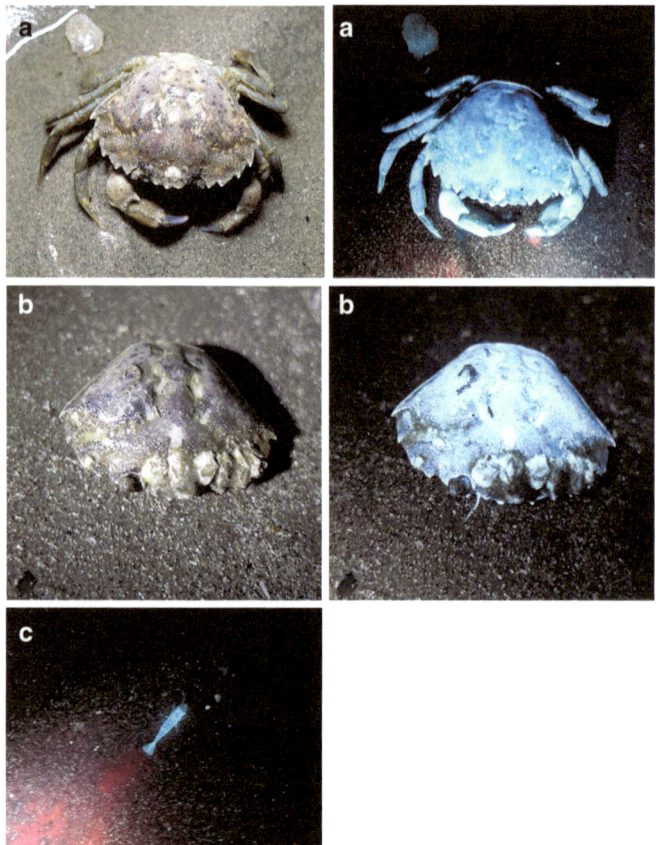

Fig. 6.7 **a** A washed-up, dead swimming crab and **b** a molted carapace of a swimming crab, each illuminated with white light (left) and with UV light (right). **c** A North Sea shrimp in UV light

on the presence of chitin, which is made up of thousands of acetylglucosamine units, which emit fluorescence light with a wavelength of about 430–440 nm, i.e., blue light [12, 13]. However, I have found that not all crab shells found glowed in UV light. Only about every third cuticle granted me this honor. Presumably, the proportion of chitin in the shell was too low.

6.2.3 Experiment: Invisible Color Diversity (North Sea)

You will need

- North Sea beach
- UV flashlight (365 nm)
- LED lamp
- Some courage to walk through the sandbanks in pitch darkness

How it works

While shining around with the white LED lamp in a sandbank, I stumbled upon a red-violet algae (Fig. 6.8a). Under UV light, this piece of seaweed fluoresces in a beautiful orange (Fig. 6.8a). What surprised me the most, however, was the sudden visibility of other algae that I had not noticed at all in the white light. Only through UV irradiation did the gray mud puddle reveal itself as a colorful collection of many delicate structures (Fig. 6.8a).

What's the science behind it?

Sea moss *(Sertularia cupressina)*, also known as cypress moss, is neither moss nor algae nor seaweed, but is a cnidarian and belongs to the hydrozoans [14, 15]. This bushy marine animal consists of a body shell made of chitin with many multiply branched chitin side branches

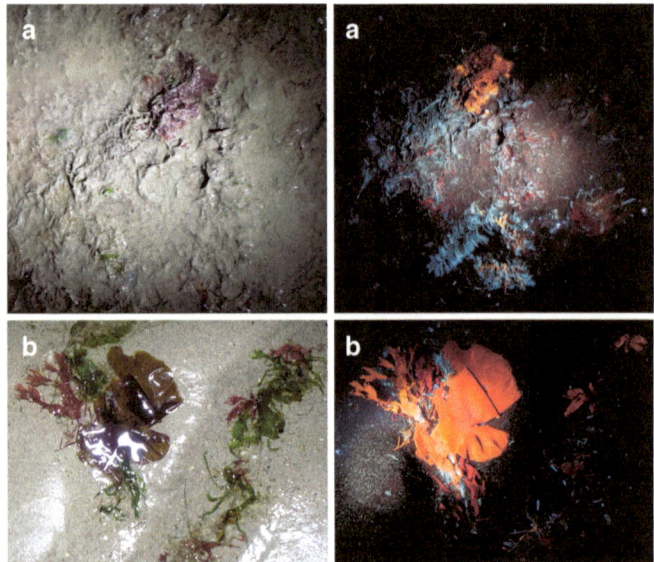

Fig. 6.8 **a** Inconspicuous mud with a piece of red algae in a sandbank illuminated with white light (left) and with UV light (right). Delicate sea moss fluoresces blue, the red algae glows orange, and smaller pieces of the green gut algae appear in bright red. **b** Algae mix of brown algae, red and green algae in a wave furrow illuminated with white light (left) and with UV light (right)

that wind spirally around the trunk. At the end of these branches sit tiny tentacle beads that catch animal and plant plankton with their stinging threads. Normally, the cypress moss is firmly anchored on the seabed, but the trawl nets of the crab fishermen tear them off and the sea current carries them to the wash margin. Due to their pale beige or gray color, they often lie inconspicuously in the mudflat, but the UV light makes them visible as blue "palm fronds".

The red-violet piece of "seaweed" in Fig. 6.8a is the Red Oak Seaweed *(Phycodrys rubens/Phyllophora rubens)*

[16]. Figure 6.8b shows an algae mix of a brown algae (Palm Seaweed, *Laminaria hyperborea*), a red algae (Red Oak Seaweed, *Phycodrys rubens/Phyllophora rubens*), two green algae species (Sea Lettuce, *Ulva lactuca* and Common Gutweed, *Ulva intestinalis*) as well as the sea moss (*Sertularia cupressina*). Red algae *(Rhodophytes)* are represented worldwide in all seas and waters with around 6000 species and contain red pigments from the class of phycobilins, such as phycoerythrin [17, 18]. The red pigments are photosynthetic dyes and help the algae collect light underwater by absorbing blue and green light in the light-harvesting complex and passing the energy on to the reaction center of the photosynthetic apparatus [19]. The deeper the water, the less light reaches the algae. The low-energy red light, which is mainly absorbed by the green chlorophyll, has only a small "immersion depth", while the significantly more energy-rich green and blue light can penetrate very far into the water. In the crystal-clear Caribbean, an algae species has been discovered at an astonishing depth of 260 m [17]. The red pigments, such as phycoerythrin, can fluoresce at a wavelength of about 580–590 nm, i.e., orange [20, 21]. The chlorophyll of the green algae shows the typical red fluorescence.

Background
No red algae, no sushi
The red algae *Porphyra* is mainly cultivated in shallow coastal waters on nets in Japan and dried into wafer-thin sheets after harvest. Known worldwide as "Nori", these mostly dark red-violet "films" are used as a mineral-rich wrapping for sushi dishes [18]. In Japan alone, around 400,000 tons of red algae are processed as food each year. Nori sheets are rich in vitamins and have a very high protein content of 30–50% [22].

6.2.4 Experiment: Bright Yellow (Baltic Sea & North Sea)

You will need

- Baltic Sea beach or North Sea beach
- UV flashlight (365 nm)
- LED lamp

How it works

Look for reddish algae at the beach's wash margin. Under UV light, they glow spectacularly in bright yellow-orange. Figure 6.9a shows the Blood Red Sea Sorrel, a beautiful but sensitive red algae, whose leaves look like feathers. In white light, the fine fibers are hardly visible to the naked eye. With UV rays, however, the delicate structure becomes clearly visible. In the North Sea wash margin, I also came across a mud algae (Fig. 6.9b) and the Red Oak Tang (Fig. 6.9c).

What's the science behind it?

Blood Red Sea Sorrel *(Delessaria sanguinea)* in Fig. 6.9a, the Worm-leaved Mud Algae *(Gracilaria vermiculophylla)* in Fig. 6.9b and the Red Oak Tang *(Phycodrys rubens/ Phyllophora rubens)* in Fig. 6.9c are all red algae, whose color pigments come from the class of phycobilins [23]. When exposed to UV radiation, they fluoresce in beautiful yellow-orange with a wavelength of about 580–590 nm [19–21].

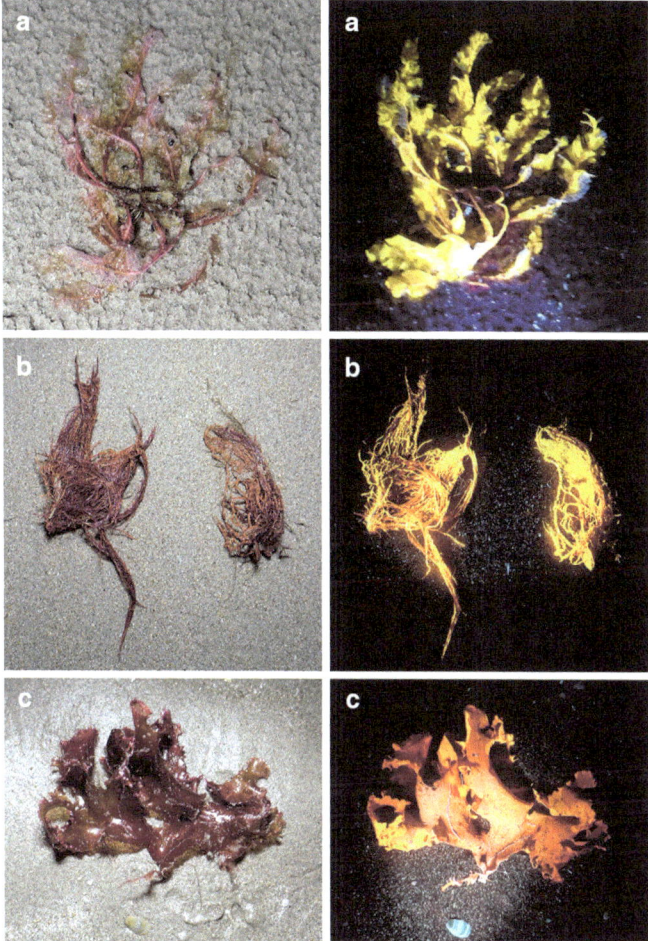

Fig. 6.9 Red algae in the wash margin of the North and Baltic Sea, illuminated with white light (left) and with UV light (right). **a** Blood Red Sea Sorrel on the Baltic Sea beach on Fischland-Darß. **b** Worm-leaved mud algae on the Belgian North Sea. **c** Red Oak Tang on the Belgian North Sea

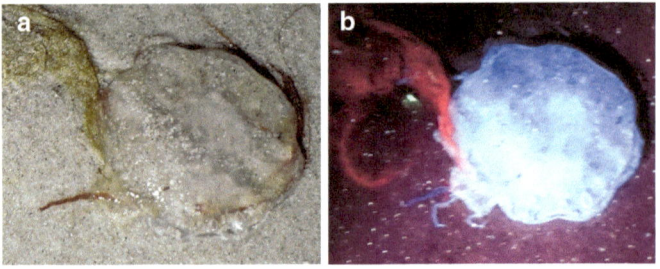

Fig. 6.10 Jellyfish and green algae, illuminated **a** with white light and **b** with UV light

6.2.5 Blue Jellyfish (Baltic Sea)

If you come across a jellyfish on the beach, you will find that some of these cnidarians also fluoresce under UV light. Figure 6.10 shows a stranded jellyfish on Fischland.

6.2.6 Baby Blue

You have to look closely to discover it—a dead baby crab, a mini beach crab on a small stone. Under UV light, their chitin shell fluoresces noticeably blue [12, 13]. In Fig. 6.11, you can see a Baltic flat shell (also known as "Red Bean") at the bottom right, which glows reddish-violet.

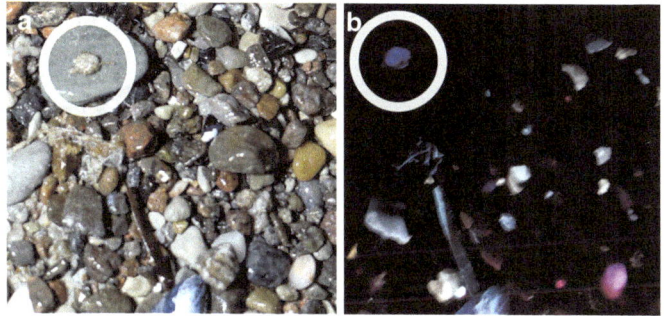

Fig. 6.11 Baby beach crab (circled) at the wash margin on Fischland-Darß illuminated **a** with white light and **b** with UV light. At the bottom right, a red Baltic flat shell is fluorescing

6.2.7 Trash on the Beach

While UV-beaming around the beach of Lubmin in the evening, a glaring orange glow caught my eye from afar. Full of anticipation of having discovered a special algae, shell, or snail, the object turned out to be a man-made plastic particle, painted with orange neon paint. A lousy piece of microplastic (approx. 5 mm) and right next to it the remains of a balloon (Fig. 6.12). The pollution of our seas already starts with such small pieces of trash. Very unfortunate.

Fig. 6.12 Waste on the Baltic Sea beach on Fischland-Darß, illuminated **a** with white light and **b** with UV light: The upper part of a balloon and a neon-colored microplastic piece

6.3   Baltic Gold and Chicken Gods

During my excursions to the Baltic Sea beaches, I have repeatedly observed vacationers on the beach who walk slowly through the rubble, bent forward, apparently in search of something valuable. If you ask these collectors, you mainly hear about two "treasures" that are worth being found: amber and chicken gods. I then also went on the occasional search for these stones on Fischland-Darß for two days and actually found something after a few hours of beach walking. Perhaps such a "treasure hunt" will make a beach walk more attractive for your children.

6.3.1 I find what you don't see—Special Stones on the Baltic Sea

Upon closer inspection, bending down and rummaging in the rubble, one can find numerous chalk microfossils, i.e., petrifications of ancient marine animals, especially near chalk cliffs, which date from the Cretaceous period (145–65 million years ago) [24, 25]. Fossil sea urchins are quite common. Chalk stones were formed about 70 million years ago at the end of the Cretaceous period from marine deposits of animal calcium shells. It is estimated that about one million years were needed for a 35 m thick layer of chalk [25]. The most spectacular and well-known chalk cliffs are located on Rügen and on the east side of the Danish island of Møn [26], where, among other things, the fantastic science museum "GeoCenter Møns Klint" offers exciting activities [27]. Figure 6.13 shows the petrifications of a small sea urchin *(Echinites)* and two brachiopods *(Brachiapoda),* which I found on Fischland-Darß.

The most common beach stone on the Baltic Sea is the dark, often black and white chalk flint, with the white area

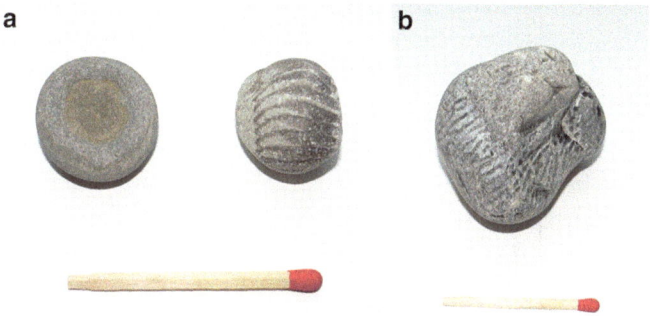

Fig. 6.13 **a** Petrified sea urchin, diameter: 2 cm (left) and petrified brachiopod, size: 2 cm (right). **b** Petrified brachiopod, size: 5 cm. The matchstick serves as a size comparison

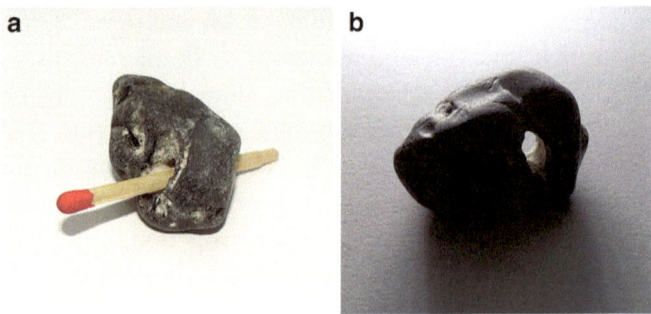

Fig. 6.14 A small chicken god flint, size: 3 cm. **a** The matchstick serves as a size comparison and as proof that there is indeed a hole in the flint. **b** The same chicken god lit from behind

on the surface being the actual chalk. The chalk consists mainly of the roundish calcium shells of the dinoflagellates and is a sedimentary or deposit rock. The flint itself is black or brown [25, 28]. It belongs to the so-called concretions, which formed in the chalk through dissolution and re-precipitation of silicon dioxide (SiO_2) and form irregularly shaped nodules. Chalk = sedimentary rock = calcium carbonate, flint = silicon dioxide from silicic acid. Highly prized among collectors are perforated flints, so-called "chicken gods", which are considered lucky charms [25, 29]. A small specimen is shown in Fig. 6.14. The formation of the holes is explained as follows: During the formation of the flint (concretion), the already petrified, calcium-rich marine animals were enclosed in the chalk and washed out by surf and weathering, as their material (calcium carbonate) was softer than the very hard flint (silicon dioxide).

6.3.2 Amber, the Gold of the Baltic Sea

Amber is most often found in driftwood after a strong autumn storm, which is washed up on the beach along with small pieces of wood due to similar densities. Promising locations for finding amber are Usedom, Hiddensee, and Fischland-Darß-Zingst [30]. Amber originated from the resin of coniferous trees that grew during the Tertiary period (65–2.6 million years ago) about 25–40 million years ago. Resin leaking from branches or the tree trunk contained microscopic air bubbles. Gradual solidification due to solar radiation caused the resin to warm up and thus the gas to escape. The resin thus became transparent and clear—this is referred to as "natural clarification" [30]. The more heat, the clearer, but also the browner the resin and thus the amber became. You are probably familiar with those magnificent specimens from movies and television, preferably with ancient insects trapped inside, as in the blockbuster "Jurassic Park". In 2023, the largest prehistoric flower ever trapped in amber, nearly 3 cm in diameter, was examined to find clues about the climate at that time [31]. The less sunlight, the less heat, the cloudier an amber. Resins running down the trunk were therefore clear, while resins that dripped onto the ground remained cloudy and milky. Over the millions of years, the resin petrified and was eventually washed into the sea by rivers. I actually found a tiny piece of amber and I am very proud of it (Fig. 6.15).

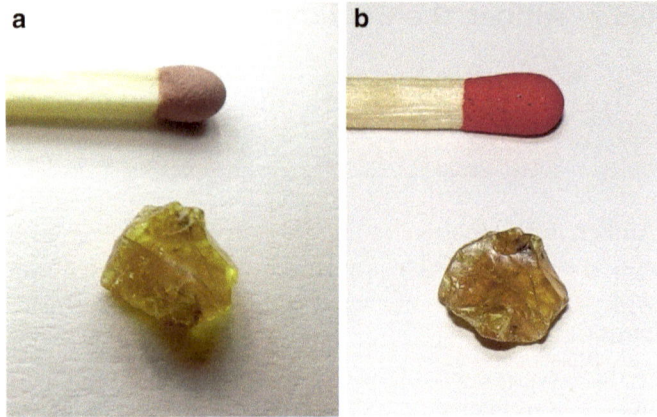

Fig. 6.15 a, b Pieces of amber, partly cloudy due to the inclusion of microscopic bubbles. Size: 0.6 cm. Found at: Fischland-Darß

To check whether it is actually petrified resin or just an ordinary stone, the best method is the float test. For this, you prepare a salt solution with two heaped tablespoons of cooking salt (30 g) in a glass of water (200 mL). Once the salt is dissolved, you put the suspected amber into the 15% salt water. If the stone floats, it is amber. If it sinks, it is just a stone. Due to its lower density, amber floats on the brine. Density of amber: approx. 1.07 g/cm^3. Density of 15% salt solution: around 1.1 g/cm^3. Warning: Yellowish chunks of white phosphorus, which look very similar to amber, are repeatedly washed up on the beaches of the German Baltic Sea. White phosphorus is highly toxic and can ignite spontaneously in the air once it has dried. It comes from incendiary bombs dropped by British air raids during World War II.

6.4 Sand Avalanches

Even as a child, I enjoyed "sand avalanche digging" because it was totally simple and exciting. A "natural disaster" in miniature without human casualties.

6.4.1 Experiment: Digging a Sand Avalanche

You will need

- Small dune on the North or Baltic Sea
- Playful spirit

How it works

Find a small dune hill and dig a little sand downwards on the "slope" with your fingers. It works particularly well if it rained the night before and the surface sand is still a bit damp. The upper (wet) sand crust remains standing while digging, while the dry, fine sand underneath goes off in an avalanche-like manner. For a short time, the mini sand avalanche "runs" by itself. A video of this can be found under Fig. 6.16.

What's the science behind it?

Sand mainly consists of the two most common elements in the earth's crust, silicon and oxygen [32]. The silicon atoms are linked with four oxygen atoms each in the form of tetrahedra to form a three-dimensional, regular crystal lattice. The silicon atom sits in the center of the tetrahedron and the four oxygen atoms occupy the four corners.

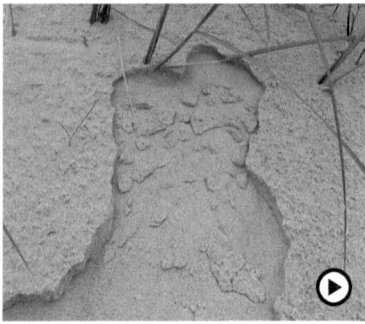

Fig. 6.16 Mini sand avalanche on the Baltic Sea beach: The video shows what happens when the angle of friction becomes larger and the frictional force becomes smaller URL: ▸ https://doi.org/10.1007/000-a6f

Each SiO_4 tetrahedron is connected to four other SiO_4 tetrahedra via their common oxygen atoms [33].

Moist sand can be easily shaped, allowing for the construction of sandcastles and even spectacular sand figures. The two hydrogen atoms of water (H_2O) are somewhat "positivized" and can interact with the somewhat "negativized" oxygen atoms of the sand (SiO_2 or SiO_4). They—figuratively speaking—shake hands. The chemist rather speaks of hydrogen bonding. Each grain of sand can "hold hands" with water molecules, thus making the entire sand stick together. Figure 6.17 illustrates this in a clear graphic.

Wet sand can be shaped, cut, scratched, etc. It sticks to hands, legs, the shovel, and in the mold. This is a case of adhesion. Especially after rain, in fog, or high humidity, the upper layer of sand is usually somewhat moist. Therefore, the upper layer of the sand avalanche holds together for a certain time due to adhesive forces. The underlying dry sand is largely water-free and very pourable. The force that holds these dry sand grains together on a dune hill is the frictional force, more precisely, the static

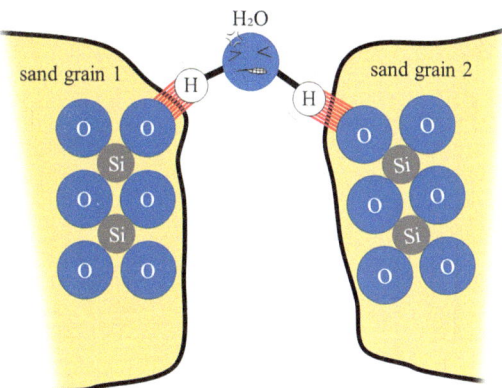

Fig. 6.17 Adhesion or: How water holds the sand grains together. H = hydrogen atom (white), O = oxygen atom (blue), Si = silicon atom (gray). Graphic: Melvin Müller

frictional force [34]. Due to their rough surface, the grains interlock and rub against each other. In the sand hill, the frictional force opposes the normal force acting perpendicular to the contact surface. In a horizontal position, the normal force equals the gravitational force or weight force pulling towards the center of the earth [34]. Important in this consideration is the all-decisive angle of friction. Before the sides of the sand hill reach an inclination angle of 45°, the sand begins to slide [34]. This is exactly what happens in this experiment. By "digging away" the sand lying below, the angle of friction becomes larger (steeper) until the critical value is exceeded and a small sand avalanche detaches.

This phenomenon can also be observed with large, piled-up sand hills on construction sites. A sand hill— no matter how large and high it is—will always have an inclination angle of less than 45°, otherwise the sand will immediately start to slide. You will never see steep, pointed sand hills.

Background

Adhesion
Attraction or interaction between particles of *different* substances. Examples: adhesive and the material to be glued; postage stamp on the licked finger; solder joint; wet leaves on the ground; mortar and brick.

Cohesion
Attraction or interaction between particles of *one* substance. Examples: water/water droplets—the water molecules in the water attract each other and form hydrogen bonds; second coat of paint on the same color; pulling on a rope.

The typical grain size of the sand of the North and Baltic Seas is between 0.06 and 2 mm.

6.5   Beyond the Horizon … it Continues …

What Udo Lindenberg mumbles so listenable and typically in his hit from 1986 is self-evident for us today. However, in those dark times of Galileo Galilei, one was threatened with death by fire or at least with ostracism or prohibition for this. In the last experiment of my book, our gaze should wander over the wide sea to the horizon. Whether you are relaxing your soul on the Mediterranean, the North Sea or the Baltic Sea—the vastness of the sea is always deeply moving to me.

6.5.1 Experiment: The Sea … Infinite Vastness … or is it?

We are 60 m high on the viewing platform of the "New Lighthouse" on Borkum (the lighthouse in the middle of the center with the red "cap"). From up there, you have a magnificent view over the entire island and the beautiful Wadden Sea. On the horizon, the summer sun sets with splendor and glory. Only the numerous massive wind turbines of the 15, 37, or 45 km distant offshore wind farms "Riffgat" and "Riffgrund I and II" impair the unobstructed view of the red fireball. But how far can you actually see? How far is the horizon, due to the spherical curvature of the earth, away? This can be easily calculated. If your children ever pester and annoyingly ask why you need math in life, you will find an answer here: Horizon expansion in the truest sense of the word with the Pythagorean theorem.

You will need

- Foresight (e.g., standing on the beach or on a lighthouse)
- Yardstick or measuring tape
- Calculator (in the smartphone)
- A bit of math

How it works

With the so-called "horizon formula", you can easily calculate the horizon distance s in kilometers [35]. It is $s = 3.57 \cdot \sqrt{h}$. Here, h stands for the eye level of the horizon viewer and is given in meters. On the viewing platform of the lighthouse, the height $h = 60$ m. Inserted into the formula, s results in a length of about 27.7 km.

Measure your eye level and/or that of your children with a yardstick or measuring tape. My viewing height is 1.75 m. Inserted into the formula, I get a good 4.7 km visibility to the horizon. And you?

What's the science behind it?

The "popular" Pythagorean theorem makes it possible. The mathematical basics are to be made clear using the graphic in Fig. 6.18.

In a right-angled triangle, the basic Eq. 6.1 applies, as the side opposite the right angle ($c =$ hypotenuse) is related to the other two sides (a and $b =$ catheti).

$$a^2 + b^2 = c^2 \tag{6.1}$$

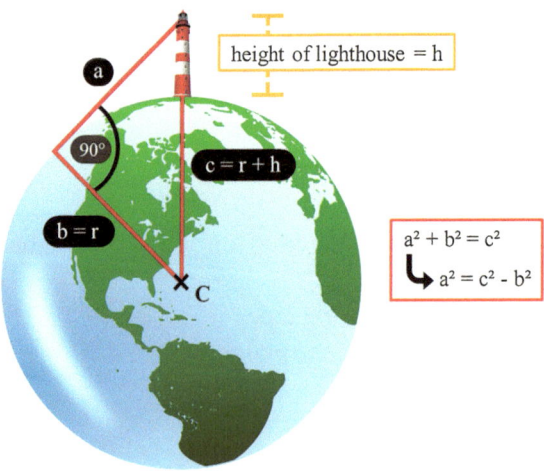

Fig. 6.18 Graphic to illustrate the horizon width using the Pythagorean theorem: $a =$ the distance to be determined from the location (lighthouse) to the horizon meets the surface tangentially perpendicular to the earth radius. b corresponds to the earth radius $r = 6371$ km. c is composed of the earth radius r plus the (eye) height of the viewer (here: at 60 m height on the lighthouse). M is the center of the earth (after [36, 37]). Graphic: Melvin Müller

The length a is sought. Therefore, the equation is solved for a (Eq. 6.2):

$$a^2 = c^2 - b^2 \qquad (6.2)$$

It applies: b corresponds to the earth radius r, which is chosen here as an average value of 6371 km. So: $b = r = 6371$ km

The length of the hypotenuse c is composed of the Earth's radius r and the (eye) height h of the location. This leads to Eq. 6.3, which, by the way, also leads to the aforementioned "horizon formula" via the binomial theorem.

$$a^2 = (r + h)^2 - r^2 \qquad (6.3)$$

In the present case, the height h of the viewing platform of the Borkum lighthouse is 60 m. So: $c = r + h = 6371$ km $+ 60$ m. Now the values are inserted, whereby the 60 m must be converted into 0.06 km. This results in the following calculation (Eq. 6.4) and after taking the root, the result in km.

$$a^2 = (6371.06 \text{ km})^2 - (6371 \text{ km})^2$$
$$a = \sqrt{6371.06^2 - 6371^2}$$
$$a = \sqrt{764.5236} \qquad (6.4)$$
$$a = 27.7 \text{ km}$$

With an adult eye height of 1.70 m ($= 0.0017$ km), the horizon visibility is about 4.7 km. Children with an eye height of, for example, 1.20 m can see 3.9 km. The sight distance of the common shore crab with a presumed eye height of 2 cm amounts to a meager 500 m. That must be enough for them.

References

1. E. Breitmaier und G. Jung, *Organische Chemie*, 7., vollständig überarbeitete und erweiterte Aufl., Georg Thieme Verlag, Stuttgart, **2012**, pp. 570–572.

2. a) S. Schramm und D. Weiß, *Biolumineszenz: Das bunte Leuchten der Natur und dessen chemische Mechanismen, Teil 1: Terrestrische Biolumineszenz*, Chem. Unserer Zeit, 57, **2023**, pp. 6–19. b) S. Schramm und D. Weiß, *Biolumineszenz: Das bunte Leuchten der Natur und dessen chemische Mechanismen, Teil 2: Maritime Biolumineszenz*, Chem. Unserer Zeit, 57, **2023**, pp. 148–161.

3. V. Storch und U. Welsch, *Kurzes Lehrbuch der Zoologie*, 8., neu bearbeitete Aufl., Springer Spektrum Verlag, Berlin Heidelberg, **2012**, pp. 441–443.

4. https://www.neuharlingersiel.de/entdecken/erlebnisse-in-neuharlingersiel/meeresleuchten/ (last access: Oct 18, 2024).

5. https://www.nationalparkhaus-wattenmeer.de/nationalpark-haus-wurster-nordseekueste/archivfacebook (last access: Oct 18, 2024).

6. https://www.domgymnasium-magdeburg.de/de/aktuelles/news/eintrag/nawi-woche-2022-auf-hallig-hooge/ (last access: Oct 18, 2024).

7. R. Reinicke, *Funde am Ostseestrand*, 2th edn., Demmler Verlag, Ribnitz-Damgarten, **2011**, p. 43.

8. T. Katoh, U. Nagashima and M. Mimuro, *Fluorescence properties of the allenic carotenoid fucoxanthin: Implication for energy transfer in photosynthetic pigment systems*, Photosynth. Res., 27, **1991**, pp. 221–226.

9. G. Quedens, *Strand und Wattenmeer*, 10., überarbeitete Aufl., BLV Buchverlag, München, **2013**, pp. 60–65.

10. V. Storch und U. Welsch, *Kurzes Lehrbuch der Zoologie*, 8., neu bearbeitete Aufl., Springer Spektrum Verlag, Berlin Heidelberg, **2012**, pp. 87–88.

11. L. Urry, M. Cain, S. Wasserman, P. Minorsky und J. Reece, M., *Campbell Biologie*, 11., aktualisierte Aufl., Pearson Verlag Deutschland, München, **2019**, p. 102.

12. M. D. Rabasovic, D. V. Pantelic, B. M. Jelenkovic et al., *Nonlinear microscopy of chitin and chitinous structures: a case study of two cave-dwelling insects*, J. Biomed. Opt., 20, **2015**, pp. 016010–1–016010–10.

13. Q. Dong, W. Qiu, L. Li, N. Tao et al., *Extraction of chitin from white shrimp (Penaeus vannamei) shells using binary ionic liquid mixtures*, J. Ind. Eng. Chem., **2023**, in press.

14. V. Storch und U. Welsch, *Kurzes Lehrbuch der Zoologie*, 8., neu bearbeitete Aufl., Springer Spektrum Verlag, Berlin Heidelberg, **2012**, pp. 460–462.

15. G. Quedens, *Strand und Wattenmeer*, 10., überarbeitete Aufl., BLV Buchverlag, München, **2013**, p. 38.

16. G. Quedens, *Strand und Wattenmeer*, 10., überarbeitete Aufl., BLV Buchverlag, München, **2013**, p. 36.

17. L. Urry, M. Cain, S. Wasserman, P. Minorsky und J. Reece, M., *Campbell Biologie*, 11. aktualisierte Aufl., Pearson Verlag Deutschland, München, **2019**, pp. 806–807.

18. J. L. Slonczewski und J. W. Foster, *Mikrobiologie*, 2th edn., Springer Spektrum Verlag, Berlin Heidelberg, **2012**, pp. 894–895.

19. H. Bannwarth, B. P. Kremer und A. Schulz, *Basiswissen Physik, Chemie und Biochemie*, 4., aktualisierte Aufl., Springer Spektrum Verlag, Berlin, **2019**, pp. 412–414.

20. C. S. French and V. K. Young, *The fluorescence spectra of red algae and the transfer of energy from phycoerythrin to phycocyanin and chlorophyll*, J. Gen. Physiol. 35, **1952**, pp. 873–890.

21. M. Beutler, K. H. Wiltshire, C. Reineke and U.-P. Hansen, *Algorithms and practical fluorescence models of the photosynthetic apparatus of red cyanobacteria and Cryptophyta designed for the fluorescence detection of red cyanobacteria and cryptophytes*, Aquat. Microb. Ecol. 35, **2004**, pp. 115–129.

22. https://de.wikipedia.org/wiki/Nori (last access: Oct 18, 2024).

23. R. Reinicke, *Funde am Ostseestrand*, 2th edn., Demmler Verlag, Ribnitz-Damgarten, **2011**, p. 48.

24. R. Reinicke, *Steine am Ostseestrand*, 7th edn., Demmler Verlag, Ribnitz-Damgarten, **2021**, pp. 36–55.

25. D. Sicker, *Kreide und Feuerstein – ungleiche Geschwister*, Chem. Unserer Zeit, 56, **2022**, pp. 197–202.

26. R. Reinicke, *Steine am Ostseestrand*, 7th edn., Demmler Verlag, Ribnitz-Damgarten, **2021**, p. 29.

27. https://moensklint.dk/de?lang=de (last access: Oct 18, 2024).

28. R. Reinicke, *Steine am Ostseestrand*, 7th edn., Demmler Verlag, Ribnitz-Damgarten, **2021**, pp. 24–28.

29. R. Reinicke, Steine am Ostseestrand, 7th edn., Demmler Verlag, Ribnitz-Damgarten, **2021**, pp. 32–33.

30. R. Reinicke, Steine am Ostseestrand, 7th edn., Demmler Verlag, Ribnitz-Damgarten, **2021**, 62–69.

31. E.-M. Sadowski and C.-C. Hofmann, *The largest amber-preserved flower revisited*, Sci. Rep., 13, **2023**, pp. 1–11.

32. G. Markl, *Minerale und Gesteine*, 3rd edn., Springer Spektrum Verlag, Berlin Heidelberg, **2015**, p. 440.

33. Holleman/Wiberg, *Anorganische Chemie*, 103rd edn., Walter de Gruyter Verlag, Berlin, **2017**, pp. 1099–1102.

34. P. A. Tipler und G. Mosca, *Physik*, 8., korrigierte und erweiterte Aufl. (Hrsg.: P. Kersten und J. Wagner), Springer Spektrum Verlag, Berlin, **2019**, pp. 118–122.

35. https://www.rhetos.de/html/lex/horizontformel.htm (last access: Oct 18, 2024).

36. https://www.bretagne-tip.de/horizont-des-meeres/berechnung.php (last access: Oct 18, 2024).

37. B. P. Kremer, *Vom Strandkorb aus betrachtet*, 1st edn., Springer Verlag, Berlin, **2021**, pp. 119–122.

Glossary

Absorption Absorption of light energy by a colored body. When absorbing light energy, the body heats up and radiates heat in the form of infrared light. Black car bodies become much hotter in sunlight than white or silver cars. Transparent bodies do not absorb light.

Acetylglucosamine A variant of glucose (dextrose), in which an OH group has been replaced by -CH_3, oxygen (O), and nitrogen (N). This simple sugar is, among other things, the building block of chitin, which consists of thousands of acetylglucosamine units and is quite similar in structure to cellulose.

Accessory Pigments Natural, plant-based photosynthetic dyes, such as carotenoids or xanthophylls, which can collect, absorb, and transfer energy to the chlorophyll in the photosynthesis apparatus. They are also referred to as auxiliary pigments.

Acid Chemically speaking, acids consist of a negatively charged non-metal group and a positively charged H-atom, called a proton (H^+). They are referred to as proton donors because acids like to give away their protons (H^+ ions). Bases are the "counterpart" to acids, they take up the H^+ ions and form water with their OH^- ions.

Adhesion Attraction or interaction between particles of *different* substances. Examples: adhesive and the material to be glued; postage stamp on a licked finger; solder joint; wet leaves on the ground; mortar and brick.

Air quality Air quality. Purity of the breathing air. The air quality is constantly checked by regular measurements.

Albedo Reflectivity of the surface of a body, an object, or a natural area. The more light is reflected, the higher the albedo effect of the respective body. An albedo value of 0.4 means: 40% of the incident sunlight is reflected and thrown back, and 60% of the light is absorbed by the body. Examples: white ice surfaces or clouds have an albedo value of 0.6–0.9, green areas about 0.15–0.2, sand and desert areas 0.3–0.6, and oceans only about 0.1.

Amino acids Building blocks of proteins. There are 20 proteinogenic, i.e., occurring in natural proteins, amino acids. They form the basis of all life, as all proteins essentially do everything: cell construction, enzymes, digestion, photosynthesis, etc.

Anthocyanins Water-soluble, natural dyes from the plant world. The name comes from Greek and means "blue flower". Almost all red, violet, or blue flowers, leaves, fruits, or shells contain anthocyanins, such as blueberries, blackberries, elderberries, raspberries, strawberries, plums, pomegranates, red cabbage, radishes, geraniums, hydrangeas, hibiscus, and many more.

Anthropogenic Man-made, caused by humans. Particularly in relation to climate change.

Anticancerogen Substances that prevent or delay the formation of cancer are referred to as anticancerogen.

Antifungal Substances that act against fungi and kill them are referred to as antifungal.

Amber Petrified resin from coniferous trees. The popular fossil resin is mainly found in the Baltic Sea region. Most ambers were formed in the Tertiary period about 55 million years ago.

Aromatic For many people, it means "fragrant" or "tasty". For the chemist, it involves a ring of six carbon atoms that looks like a honeycomb and has special chemical properties. The free electrons of all six carbon atoms are evenly distributed over the entire ring and form a kind of unity, an electron cloud.

Arthropodin A wild mixture of hundreds of different structural proteins, which together with chitin forms the hard exoskeleton of arthropods and crustaceans.

Atmosphere The gaseous envelope of the Earth, consisting of nitrogen (approx. 78%), oxygen (about 21%), argon (approx. 0.9%) as well as trace gases such as water vapor, ozone, carbon dioxide, methane, nitrous oxide, etc. The atmosphere is divided into the troposphere, stratosphere, mesosphere, thermosphere, and exosphere, which begins at about 800 km altitude. The ozone layer, which primarily absorbs the harmful UV-C and parts of the UV-B radiation, is located in the stratosphere. The International Space Station (ISS) orbits in the thermosphere at an altitude of around 400 km.

Atom Smallest particles of matter that cannot be further broken down without destroying them. Although the Greek word atom means "indivisible", we now know from precise measurements at particle accelerators that atomic nuclei themselves consist of even smaller particles, the quarks, which are the actual elementary particles.

ATP Adenosine triphosphate. *The* universal energy carrier of all life. It consists of a sugar (ribose), a nitrogen base (adenine), and three phosphate units. When a phosphate unit is detached, the chemically bound energy is released and can be used by the cells, for example, for the synthesis of glucose from carbon dioxide in photosynthesis.

Bacteria Microscopically small single-celled organisms as well as prokaryotes that do not have a cell nucleus. Bacteria are water dwellers. Without water or moisture, there are no bacteria. Therefore, water-depleting substances such as salt and sugar or drying make food durable.

Bacteriostatic Substances that act against bacteria and inhibit their reproduction are referred to as bacteriostatic.

Base Chemically speaking, bases consist of a positively charged metal ion and a negatively charged OH group ($Me^+ + OH^-$). They are referred to as proton catchers because their hydroxide (OH^-) group can absorb protons (H^+ ions) from acids and thus react to form water. Bases are the "counterpart" to acids. Well-known bases include, for example, sodium hydroxide and potassium hydroxide.

Bimetal A metal strip consisting of two different, firmly connected metal layers, for example brass and steel. When the temperature increases, the different degrees of thermal expansion of both metals cause a bend upwards or downwards. This effect is used for temperature control, among other things, in heaters, irons, kettles, toasters, boilers, and coffee machines.

Bioluminescence Enzymatically catalyzed chemiluminescence ("cold light") in living organisms (e.g., fireflies, marine bioluminescent organisms).

Biopolymer Long chains or gigantic molecules that are produced in living beings, thus are of natural origin. Examples: Cellulose, Chitin, Proteins, Collagen, Starch.

Blue-green algae Are not algae, but cyanobacteria. They perform photosynthesis with both chlorophyll and blue light-collecting pigments (e.g., phycocyanin) and have been inhabiting the Earth for 3.5 billion years. Some species use hydrogen sulfide instead of water in the photosynthesis reaction.

Brachiopods *(Brachiapoda)* Sea-dwelling creatures that resemble shells and were embedded as fossils in chalk (= sedimentary rock). They have been living on Earth for over 500 million years.

Brown algae True algae that live in the sea and are mainly found in the mudflats and on rocky coasts. They can grow up to 70 m long.

Bryology The study of mosses.

Carbohydrate Sugary substances consisting of carbon, hydrogen, and oxygen atoms. Formula: $C_n(H_2O)_n$. Known carbohydrates include: Glucose (grape sugar), Fructose (fruit sugar), Sucrose (sugar), Starch, Cellulose, Chitin, Vitamin C.

Carbon Dioxide An elongated, symmetrical molecule made up of three atoms with infrared activity. *The* greenhouse gas par excellence that is harmful to the climate. Over 36 billion tons of carbon dioxide are emitted on Earth each year. Formula: CO_2.

Carnivore Carnivorous plant.

Carotenoids Fat-soluble, water-insoluble, natural pigments from the plant world. The name is derived from the Latin for "carrot". The suffix "-oid" means "similar". So far, about 700 naturally occurring carotenoids are known. The "original substance", carotene, consists of 40 carbon and 56 hydrogen atoms. The well-known β-carotene, also called provitamin A, is converted into vitamin A in the body. Carotenoids are capable of collecting light and are significantly involved in photosynthesis as accessory pigments (= auxiliary pigments). They also protect plants from excessive UV radiation. Their color spectrum ranges from yellow to orange to red. Examples: Carotene (including carrots, pumpkin, apricots, nectarines, peaches, salads), Lycopene (tomatoes, watermelons).

Cellulose Biopolymer composed of thousands of glucose units linked into long chains (polysaccharide). It is the main component of plant cell walls and, with an "annual production" of over 100 billion tons, it is the most common biopolymer on Earth. For humans, cellulose is a significant raw material for paper production.

Chalk Formed about 67 million years ago at the end of the Cretaceous period from marine deposits of animal calcium shells.

Charge distribution Spatial distribution of electrical (positive and negative) charges in molecules.

Chemiluminescence Cold light. A luminous phenomenon produced by a chemical reaction. Examples: glow sticks, fireflies, bioluminescent marine organisms.

Chicken God Flint with a hole.

Chitin Biopolymer made up of thousands of acetylglucosamine units, which on its own is soft and flexible and resembles cellulose. Only through cross-linking hundreds of structural proteins (arthropodins) is sclerotin formed, which together with chitin becomes the hard and stable exoskeleton (cuticle) of insects. In crustaceans, calcium carbonate (lime) is additionally deposited to increase hardness and stability, similar to reinforced concrete. Chitin has been a natural product for about 500 million years and with an annual "production" of over 10 billion tons, it is the second most common biopolymer on earth. Especially the animal plankton of the world's oceans (krill) contributes significantly to this. The world champions in biopolymer production are plants (cellulose).

Chlorophyll Green leaf pigment, which is not only contained in plant cells, but also in photosynthetically active organisms such as cyanobacteria (blue-green algae), green algae, red and brown algae. Chlorophyll pigments are anchored in the thylakoid membrane inside the chloroplasts and are the main actors of photosynthesis. So far, six different chlorophyll molecules are known, with chlorophyll *a* and *b* being the most common and important representatives.

Chloroplast A small cellular compartment, organelle in plants and algae, where photosynthesis takes place. Inside, the thylakoid membranes stack up with the embedded chlorophyll and carotenoid molecules, which collect light and pass the energy on to the photosystem. The dark reaction, the so-called Calvin cycle, in which carbon dioxide is converted into glucose, on the other hand, takes place outside the thylakoid stacks in the "free" fluid (stroma) within the chloroplasts.

Ciliates Ciliates are single-celled microorganisms (zooplankton) and occur with around 7500 species in all waters of the world. With their many cilia on the body surface, they are able to swim forwards, backwards, in arcs, and at different speeds. They move extremely quickly and are very lively. Size: 10–300 µm.

Climate Neutrality No influence on the Earth's climate by humanity. The greenhouse gases that one emits must be removed from the air somewhere else. Therefore, when there is a balance between emission and absorption of carbon dioxide, we speak of climate neutrality. If energy production were achieved solely through wind and solar power (and green hydrogen), then there would be no CO_2 emissions at all, and climate neutrality would be perfect.

Cnidarian Multicellular, simply structured animals with a nucleus in their cells (eukaryotes), equipped with stinging threads for prey capture. Cnidarians include, among others, jellyfish and anemones.

CO_2-equivalent Assessment and weighting of the greenhouse potential of a gas that is not carbon dioxide. Examples: One ton of methane (CH_4) emissions is equivalent to or would be equivalent to the emission of 24 tons of carbon dioxide (CO_2). One ton of nitrous oxide (N_2O) emissions is equivalent to or would be equivalent to the emission of 300 tons of carbon dioxide (CO_2).

Coherence, coherent Phase equality in electromagnetic waves. Simply put: When light waves come out exactly in sync from a light source, for example from a laser pointer, they can perfectly overlap and become the characteristic laser beam.

Cohesion Attraction or interaction between particles of *one* substance. Examples: Water/water droplets—the water molecules in water attract each other and form hydrogen bonds; second coat of paint on the same color; Pulling on a rope.

Contact Angle Also called wetting angle. The angle formed by the surface of a droplet with the solid surface on which the droplet rests.

Convection Flow transport.

Coumarin A natural, aromatic plant substance with a characteristic smell of woodruff. Chemically, coumarin belongs to the benzopyrones. Derivatives of coumarin are mainly used as fragrances in perfumery or as anticoagulant medications (Marcumar®, Falithrom®) for thrombosis prophylaxis.

Curette A small, sharp metal ring mounted on a rod, which can be used to perform scrapings or excisions. Skin curettes, for example, are used for the removal of warts.

Cuticula Exoskeleton in insects and other arthropods consisting of chitin and cross-linked structural proteins to form a stable and hard "outer skin". In plants, the cuticula is a waxy layer on the leaf surface.

Cyanobacteria Blue-green algae that carry out photosynthesis using chlorophyll and blue-violet pigments.

Crystal lattice The regular arrangement of atoms in a solid (metals, ice, salts).

Dark reaction Light-independent chemical reaction in photosynthesis. The energy molecules ATP and electron transport substances NADPH, formed by the light reaction, cause the build-up of glucose from carbon dioxide.

Diatoms They are characterized by their silicate shells and skeletons made of silicon dioxide (SiO_2) with beautiful architecture. The formation of chalk cliffs and flints is based, among other things, on these organisms.

Deformation vibration Molecular vibration in which the position of the atoms changes in at least one bond angle. Prerequisite: at least three atoms per molecule, such as water (H_2O), carbon dioxide (CO_2), laughing gas (N_2O). But not with oxygen (O_2) or nitrogen (N_2). The deformation vibration is excited by infrared radiation and plays a central role in the greenhouse effect.

Diethyl Ether *The* ether, formerly also referred to as Äther, is the most well-known etheric compound in the world of molecules. Ether is a highly volatile and highly flammable liquid with a numbing effect. In the 18th and 19th centuries, ether was used as an anesthetic and revolutionized medical surgery.

Digestive enzymes Substances that break down our food into its individual components to make them accessible for metabolism. Long-chain and large molecules are split into smaller compounds. This starts in the mouth and continues in the stomach to the intestine.

Dinoflagellates Belong to animal plankton (zooplankton) and are eukaryotic single-celled organisms, with an average size of 10–100 µm. Some species perform bioluminescence, which is triggered by mechanical stimulation, such as wave movements, wind, or other living beings. The thereby "indented" cell membrane leads to a chemical, enzymatic reaction, in which blue light with a wavelength of about 475 nm is emitted.

Dispersion, dispersive Describes the effect that different wavelengths of light or sound are refracted differently in different media. The reason for this is the different propagation speed of the various light wavelengths in a medium. The splitting of white light into its colors in a prism or in raindrops into the colors of the rainbow is such a dispersion effect. The dispersion effect occurs when transitioning from the medium of air to the medium of glass or water.

Electron One of the three main elementary particles. It carries a negative charge, which is defined as -1. The actual charge is -1.6×10^{-19} C. Its absolute mass is absurdly small, namely 9.1×10^{-28} g. The relative mass compared to the carbon atom is given as 0.0005 u. Electrons contribute virtually nothing to the weight of an atom. However, they are responsible for all chemical reactions, whether you light a match or plants perform photosynthesis. The electrons do *everything* (except for nuclear reactions). Symbol: e^-.

Emission From Latin: *emittere* = to emit. The emission can refer to car exhaust, fine dust, greenhouse gases, radiations of any kind, but also money and securities.

Enzyme Enzymes are molecular machines that carry out virtually all biochemical reactions of metabolism in all living beings, usually in the form of catalysis. Therefore, they are also referred to as "biocatalysts". With their presence, they lower the activation energy of a biochemical reaction, which then proceeds extremely much faster. Example: The decomposition of the harmful hydrogen peroxide H_2O_2 into harmless water (H_2O) and oxygen (O_2) proceeds without a catalyst at a relative reaction speed of 1. With platinum, which is also used in the exhaust catalyst, the reaction speed already increases to 800. With the enzyme "catalase", the decomposition races away at a breakneck speed of $3 \cdot 10^{11}$. Enzymes are incredibly fast machines. The "acetylcholinesterase", an important enzyme in the nervous system that breaks down the neurotransmitter (nerve messenger substance) acetylcholine into choline and acetic acid (otherwise cramps, paralysis and death threaten): One molecule of enzyme takes down *per second* around 14,000 acetylcholine molecules! Incredible! The "carbonic anhydrase" is even more extreme: It is mainly found in red blood cells and ensures that gaseous carbon dioxide (CO_2) is converted into water-soluble bicarbonate (HCO_3^-). Very important thing, otherwise we suffocate. One molecule of enzyme converts *per second* a whopping 10 million molecules of carbon dioxide into carbonic acid. World record! The "catalase" is also such a "Usain Bolt" of enzymes. You may have noticed: All enzymes end in "-ase". And: Enzymes are always proteins, but not every protein is an enzyme.

Excited State The electrons of an atom are usually in the lowest energy state, the "ground state". By absorbing energy, for example through light or heat, electrons can transition to states of higher energy. These are referred to as "excited states".

Exhaust Catalyst A purification system for the exhaust gases of internal combustion engines in motor vehicles, also briefly called a 3-way cat. In three chemical conversion processes, the three pollutants carbon monoxide (CO), nitrogen oxides (NO_x) and incompletely burned gasoline or diesel are converted into carbon dioxide (CO_2), nitrogen (N_2) and water (H_2O). Hence the name "3-way cat". The heart of the catalyst mounted in the exhaust system are porous ceramic honeycombs, which are coated with precious metals such as platinum, palladium or rhodium as active catalyst substances.

Exocarp The outer layer (shell) of a fruit from seed plants.

Erythrocytes Red blood cells. They are responsible for oxygen transport using the red blood pigment hemoglobin and convert gaseous carbon dioxide into soluble carbonic acid. Total number in the body: about 30 trillion. Number in a drop of blood: approx. 250 million. Lifespan: 4 months. Renewal rate: 2.4 million per second. Place of formation: red bone marrow.

Fasciculin Neurotoxin of the green mamba, which blocks acetylcholinesterase.

Filament A thin thread, a fiber.

Flint In the chalk, the calcareous deposit or sedimentary rock of prehistoric times, the hard, black flint was formed by the dissolution and re-precipitation of silicon dioxide (SiO_2). This process is also referred to as concretion.

Flue gas filter system Cleaning system to remove pollutants from flue gas (smoke). Harmful solids and gases are separated using catalysts, activated carbon, gas scrubbing, electro and fabric filters.

Fluorescence Light emission upon light irradiation. Light of a certain wavelength is absorbed, the dye molecules are excited and briefly elevated to a higher energy level for picoseconds. When "falling down" to the ground state, some energy is emitted as light. Example: highlighters, optical brighteners. But without light, there is no fluorescence: no light—no fluorescence!

Folic Acid Vitamin B_9.

Frequency Frequency, number of oscillations. The frequency indicates how often an oscillation occurs per second. Unit: 1 Hz (Hz) = 1 oscillation per second = $1 \cdot s^{-1}$.

Friction Angle The angle under which a granular material (sand) can be loaded without slipping. It corresponds to the ratio between the frictional force and the normal force on the friction surface.

Friction Force Force between two touching bodies (object on base). It is determined by the weight force and the coefficient of friction. The coefficient of friction depends on the rubbing materials and their surface texture, and one distinguishes between static, sliding, and rolling friction. The static friction is always greater than the sliding or rolling friction. Friction forces play a major role in everyday life. Examples: nail or screw in the wall, knot in a rope, walking, running, racing, car tires on road, cycling, braking of vehicles, lubricants in moving bearings and joints.

Fucoxanthin A pigment from the series of xanthophylls, the oxidized molecules of carotenoids.

Fungal spores "Seed grains" for asexual reproduction and spread of fungi, algae, mosses, and ferns. Spores are not sex cells. They are very robust and resistant and can survive for a very long time.

Gasoline A mixture of light boiling, liquid hydrocarbons with chain lengths of 5–10 carbon atoms (including pentane, hexane, octane), which is obtained by distillation from petroleum and used as fuel in internal combustion engines.

Glucose The most important simple sugar (monosaccharide) with the molecular formula $C_6H_{12}O_6$, thus containing six carbon atoms and therefore belongs to the hexoses. The ending "-ose" always indicates that it is a sugar molecule. Other names for glucose are dextrose and grape sugar.

Gravity Weight force. It causes the "heaviness" on Earth and depends on the masses of the two attracting bodies and the distance between their centers. According to the law of gravity, the weight or gravitational force $F = m \cdot g$, where g is the acceleration due to gravity or the local factor. Gravity always acts towards the center of the Earth and ensures that no person "falls off" the Earth, and that rain always trickles down.

Green Algae Belong to the Chlorophytae with over 7000 species worldwide and are represented in freshwater by 90%. Their range extends from tiny single cells to multicellular organisms the size of a hand. Green algae perform photosynthesis using chlorophyll and other light-collecting pigments.

Greenhouse Effect It describes the effect of greenhouse gases in the atmosphere on the temperature of the Earth's surface. The natural greenhouse effect brings the average temperature of the Earth to +14 °C, which without the gases water vapor, carbon dioxide, methane, and nitrous oxide would be 33 °C lower at −18 °C. This balance, millions of years old, is dramatically disturbed by the man-made (anthropogenic) greenhouse effect. The massive additional emission of the greenhouse gases carbon dioxide, methane, and nitrous oxide leads to dramatic changes in the world's climate, air, and sea temperatures.

Greenhouse gas Gases in the Earth's atmosphere that cause the greenhouse effect. Most greenhouse gases have a natural origin, but also a man-made one. The best-known greenhouse gases carbon dioxide (CO_2), methane (CH_4), and nitrous oxide (N_2O) are naturally found in low concentrations in the atmosphere. However, the proportion has drastically increased since the beginning of the last century due to various man-made sources.

Ground state The state of lowest energy, in which the electrons of an atom are located. As long as no energy is sent to atoms or molecules, all electrons remain "quiet" and essentially buzz around the atomic nucleus on the "ground floor".

Heat Capacity The ability of a body or substance to absorb and store energy in the form of heat. The liquid world record holder with the highest heat capacity worldwide is water. Therefore, the fire department likes to extinguish fires with water, and heaters use water as a heat transfer medium.

Heavy metal Metals whose density is greater than 5 g/cm^3. These are often toxic and environmentally harmful substances. Examples: Lead, Copper, Cadmium, Mercury, Uranium, Plutonium.

Hormone Messenger substances that are produced at one location in the body but exert their effect at another location and play a pivotal role in cell-to-cell communication. The main production sites are the thyroid, pancreas, adrenal gland, hypothalamus (in the brain), pituitary gland (in the brain), as well as the ovaries and testes. Hormones regulate many processes, such as fat and sugar metabolism (insulin), bone growth, muscle development, sexual development, and the menstrual cycle in women.

Huygens' Principle This basic rule explains the propagation of light waves. Every point that is hit by a light wave front becomes the origin of a small, spherical elementary wave. Huygens' Principle is of great importance in the refraction of light at the boundary of two media.

Hydrogen Bonding The two hydrogen atoms of water (H_2O) are somewhat "positivized" and can form a bond with the somewhat "negativized" oxygen atoms of neighboring water molecules.

Hydrophilic Water-loving, water-attracting, fat-repelling.

Hydrophobic Fat-loving, fat-attracting, water-repelling.

Hypholomin A specific fungal substance that can fluoresce.

Immune system Defense mechanism of an organism against foreign invaders, pathogens or disease germs such as bacteria or viruses. The highly developed immune system of mammals consists of specialized immune cells and tailor-made antibodies.

Indicator Acid-base indicators are organic dyes and indicate the concentration of hydrogen ions (H^+) through a color value. As the environment changes from acidic to neutral to alkaline (and vice versa), the structure of the dyes and thus the light absorption properties change. Anthocyanins are well-known indicators from nature (red cabbage, purple cabbage), phenolphthalein is known to some from chemistry lessons. If the pH value of a solution is 7, it is neutral. In an acidic environment, the pH value is between 1 and 7, in an alkaline environment the pH value is 7 to 14. The smaller the pH value, the more H^+ ions are in the solution. A universal indicator contains several dyes, so that the entire pH range can be represented with a corresponding color.

Infrared-active Molecules that can interact with infrared rays are referred to as IR-active. Such substances can absorb IR radiation and be stimulated to vibrate by this energy absorption. Stretching and deformation vibrations are distinguished. A prerequisite for IR activity is the presence of a dipole moment, i.e., the molecule must consist of "negativized" and "positivized" atoms. Water (H_2O) and carbon dioxide (CO_2)

are such dipole molecules. However, vibrations aligned to the center of symmetry are IR-inactive, dipole or not. The vibration must change, otherwise, there will be no absorption of the IR radiation. The absorption of IR radiation by CO_2 plays a crucial role in global warming.

Infrared light/Infrared radiation Abbreviation: IR. Electromagnetic radiation beyond visible red light, which is divided into near, mid, and far IR, with wavelengths from 780 nm to 1000 µm. Infrared light opens up numerous applications: among others, heat lamps, identification of molecules, space exploration, painting examinations, remote controls, thermal imaging cameras.

Interference pattern Superposition of light, sound, or water waves. When several waves meet, they can penetrate each other and create a wave pattern. If you throw two stones into a lake, the concentric rings overlap. If wave crest meets wave crest exactly synchronously, it is amplified. If wave crest meets wave trough exactly synchronously, both are cancelled out. In between, all possible patterns exist.

IPCC Intergovernmental Panel on Climate Change (Interstate Committee for Climate Changes, in short: World Climate Council). Founded in 1988. The committee is intended to compile, evaluate the global state of research and the scientific findings, and assess the resulting risks and impacts of global warming.

Isothiocyanate Isothiocyanates are sulfur-containing substances with the chemical formula R-N=C=S. Mustard oils also belong to the isothiocyanates, for example, the allyl mustard oil CH_2=CH-CH_2-N=C=S. They smell and taste pungent and sharp, and among other things, give mustard, radishes, horseradish, and cress their characteristic sharpness.

Isotropic Independent of the direction of view. In all directions, an isotropic substance exhibits the same chemical and physical properties.

Jochalgen Ornamental algae. They are the most beautiful green algae with delicate and often symmetrical structure, are unicellular and only occur in freshwater.

Laughing Gas Nitrous oxide, an infrared-active gas, is an elongated molecule made up of three atoms, also used as "whipping gas" in spray cans to whip cream. It occurs from natural sources in the atmosphere, but also through the use of artificial fertilizer. In medicine, it serves as an anesthetic. Laughing gas has a greenhouse potential 300 times higher than CO_2. Every year, about 3 billion tons of CO_2 equivalents are emitted on Earth. Formula: N_2O.

Leukocytes White blood cells. Immune defense cells of the body. Total number in the body: about 30–60 million; Number in a drop of blood: approx. 200,000–500,000. Lifespan: days to weeks.

Lichen Symbiotic relationship between a fungus (mycobiont) and a light-utilizing species (photobiont), such as a green alga or cyanobacteria. They grow on various substrates, on stone, on tree barks, branches, dead wood, and even on garden gates and fences. Lichens are not parasites and occur with 25,000 species all over the world.

Lichenology Study of lichens.

Lichen station Term for an area where a certain number of trees are examined for lichen growth, the lichen species are identified and counted, thus making a statement about the air quality possible.

Ligase Enzyme that links molecules together. Examples: DNA ligase—links two DNA strands together. Aminoacyl-tRNA synthetase—links an amino acid to its tRNA (important for protein biosynthesis).

Light-harvesting complex LHC. Collection of membrane proteins in the photosynthetic membranes of organisms that perform photosynthesis. In plants, the LHC is located in the thylakoid membrane of the chloroplasts. Typical LHCs contain around 200 molecules of chlorophyll and about 50 carotenoids as additional light-harvesting pigments (accessory antenna pigments). The LHC absorbs sunlight and transmits this energy through the numerous pigments at an incredible speed to the reaction center, where the light reaction of photosynthesis takes place. English: *light harvesting complex*, LHC.

Light reaction The light reaction takes place in plants in photosystems I and II. There, the absorbed light energy is converted into chemical energy. First, electrons are released by charge separation, by breaking down water into oxygen gas and hydrogen ions. Ultimately, the electrons lead to the formation of ATP and NADPH, which enter into the dark reaction in the further course.

Lignin Lignin is a large biopolymer from phenol units, bridged via oxygen atoms.

Lotus Effect Non-wettability of a surface, on which water droplets bead up spherically. In the process, all dirt particles are also taken from the surface and washed away. The prime example is the lotus flower with its self-cleaning leaves. But kohlrabi leaves also have the lotus effect.

Lye Also called a base. The counterpart to an acid. Hydrogen ion catcher. Acids release H^+ ions and lyes absorb them. Lyes have a pH value between 7 and 14.

Membrane Biological "wall".

Membrane protein Proteins that are embedded in a cell membrane and usually "stick out" at the top and protrude into the cell at the bottom.

Methane Combustible gas. Main component of natural gas. Biogas. Occurs from natural sources in the atmosphere, but primarily through massive livestock farming, rice cultivation, and natural gas and oil extraction. Methane has a greenhouse potential 28 times higher than CO_2. Approximately 8 billion tons of CO_2 equivalents are emitted on Earth each year. Formula: CH_4.

Micelle A microscopically small sphere made of molecules that exhibit both a water-loving and a fat-loving group, structured similarly to a matchstick. The "heads" align with "heads" and the "wooden sticks" with the "wooden sticks". The resulting most stable form is a sphere, with the "heads" pointing outwards and the "sticks" inwards.

Micrometer 10^{-6} m = 1 thousandth of a mm or 1 thousandth of a meter. Symbol: μm.

Microorganisms Microscopically small organisms (including bacteria, plankton, algae, fungi).

Microplastics Particles of plastic with a size from $1\,\mu m$ to $5\ mm\,(= 1\,\mu m$ to $5000\,\mu m)$.

Mineral Solids occurring in nature with a specific chemical composition and structure.

Mite Tiny arachnids (arthropods).

Molecule Substance that consists of at least two or more atoms. The hydrogen molecule consists of two atoms. Proteins are molecules with thousands of atoms.

Mycobiont A lichen fungus that enters into symbiosis with green algae or cyanobacteria.

NADP/NADPH Nicotinamide adenine dinucleotide phosphate. *The* universal substance of living nature for the transport of electrons. The nicotinamide hanging at the "top" of the molecule can take up and release two electrons (and one H^+ ion). NADPH houses two electrons, which it receives as the final runner of a relay after going through photosynthesis, and brings them to the Calvin cycle. There, the electrons are needed for the conversion of CO_2 to glucose.

Nanometer 10^{-9} m = 1 millionth of a millimeter or 1 billionth of a meter. Symbol: nm.

Nanoplastic Particles with a size of 15–1000 nm = 0.015–1 μm.

Nitric Acid The most well-known and stable oxygen acid of nitrogen. Uses: among others, for the production of fertilizers, dyes, and explosives. Its salts are called nitrates. Formula: HNO_3.

Nitrogen Dioxide Reddish-brown, toxic, pungent-smelling gas. Nitric acid is formed with water. Formula: NO_2.

Nitrogen Oxide Collective term for various gaseous compounds that are composed of nitrogen (N) and oxygen (O). The two most important compounds are nitrogen monoxide (NO) and nitrogen dioxide (NO_2). They are produced by the combustion of fossil fuels and can lead to a variety of negative health and environmental effects. Formula: NO_x.

Noctilucae Marine luminescent microorganisms. Tiny single-celled organisms (dinoflagellates) that produce bioluminescence, or "cold light," and can glow blue in the dark (marine luminescence).

Normal force This force occurs between two bodies on their contact surface and is determined, among other things, by the frictional force. If a body lies horizontally on the ground (support surface), then the normal force corresponds to the gravitational force of the body. On a sloping support surface, the normal force acts perpendicular to the support surface.

Nucleic Acid Nucleic acids, meaning cell nucleus acids, such as the universal genetic substance DNA. Nucleic acids consist of nucleotides, which in turn are each composed of a molecule of sugar (deoxyribose), a phosphate group (PO_4^-) and an organic nitrogen base (adenine, thymine, cytosine and guanine). The name "acid" is derived from phosphoric acid (H_3PO_4), whose acid residue is part of the nucleotides.

Oarfoot crab *Copepoda*. Size approx. 0.5–2 mm. About 400 species are known worldwide, oarfoot crabs are mainly found in standing waters (freshwater) in the shore area. They beat their front antennas abruptly backwards to move forward.

Ooporphyrin An important precursor for the natural production of chlorophyll or hemoglobin. Also referred to as Protoporphyrin IX.

Oral mucosa The mucous layer in the oral cavity, especially on the inner surfaces of the cheeks.

Oxidation Transfer of oxygen to a molecule. In a narrower sense: electron donation. In the oxidation of iron to rust, the iron absorbs oxygen. In this process, elemental iron is oxidized to iron ions (Fe^{3+}), thus giving away three of its electrons, which migrate to the oxygen (O_2) and make it negative to O^{2-}. Rust therefore consists of Fe_2O_3 with the typical rust-red color. The combustion of hydrogen with oxygen to water (fuel cell) is also an oxidation and electron donation. Two hydrogen atoms each give their only electron to oxygen and are oxidized. The oxygen gas (O_2) transforms by absorbing the two electrons to O^{2-} and forms water together with the two H^+ ions (protons).

Ozone A particularly reactive form of oxygen in the form of a triatomic molecule. It forms in the stratosphere at an altitude of about 50 km and shields the Earth from harmful UV-B and UV-C rays. Due to its very strong oxidizing effect, it is used as a means against germs in water (swimming pools), against exhaust gases and odors, as well as a bleaching agent. Formula: O_3.

Papilla Small, round to conical elevation on surfaces.

Parasite A living organism that survives at the expense of another organism (host) and harms the host. Examples: Mistletoe on trees, mites/fleas/lice/ticks on the skin, worms in the intestine, malaria pathogens in the blood.

Parietin A yellowish dye that fluoresces red-orange under UV light. Belongs to the chemical group of anthraquinones and is the color-giving pigment of the common yellow lichen.

PET Polyethylene terephthalate. A plastic used for making plastic bottles, textile fibers (polyester), and much more. The basic building block is terephthalic acid, a dicarboxylic acid, which is esterified with ethylene glycol (ethanediol) to form a long polymer. Worldwide annual production: 70 million tons. Partially recyclable. Since 2021, there have also been enzymatically recycled PET bottles.

pH Indicates the concentration of hydrogen ions in an aqueous solution. Since the concentrations are very small and range from 10^{-1} to 10^{-14} mol/L, the pH value was defined as the negative decimal logarithm. This saves a lot of calculation and the pH values are presented as positive numbers from 1 to 14. $pH = 7$ means neutral, i.e., there are as many H^+ ions as there are OH^- ions. With a pH value of 1–6, there are many more H^+ ions than OH^- ions, the liquid is an acid. In an alkali (base), the pH value with values between 8 and 14, on the other hand, indicates very few H^+ ions and many OH^- ions.

Phenol Hydroxybenzene, composed of a carbon six-ring ("honeycomb") with an OH group on one of the carbon atoms. An important raw material for the production of phenoplasts and epoxy resins.

Photobiont Photosynthesis-capable partner in a lichen—a green alga or a cyanobacterium.

Photochemical Chemical reactions that occur due to the influence of light.

Photoreceptor A light receiver that can convert light into bioelectrical excitation.

Photosynthesis With the help of sunlight, large amounts of carbohydrates are formed from water and carbon dioxide in green plants. Photosynthesis takes place in two partial steps: The energy conversion as a light (light dependent) reaction in the photosystems and the substance conversion as a dark (light independent) reaction from carbon dioxide to glucose.

Photosynthesis Apparatus The interplay of the two photosystems I and II during photosynthesis.

Photosynthesis pigment Light-collecting pigments that can transmit light and transfer it to other molecules. Examples: chlorophylls, carotenoids, xanthophylls. About 300 different light-collecting "antenna pigments" are known.

Photosynthesis proteins Proteins that are involved in photosynthesis.

Photosystem The light reaction takes place in the two photosystems I and II. The photosystems form the central "heart" of plant photosynthesis and consist of numerous light-harvesting complexes that group around a reaction center.

Phycobilins Light-collecting accessory pigments of photosynthesis, which occur among others in red algae and cyanobacteria (blue-green algae).

Phycoerythrin A light-collecting accessory red pigment of photosynthesis, which occurs among others in red algae and cyanobacteria (blue-green algae). Belongs to the class of phycobilins.

Phytoplankton Plant plankton, single-celled microorganisms.

Pigment Insoluble dye, also referred to as coloring agent. Examples: a) Watercolors are water-insoluble, colored pigments that are suspended with water and a brush and transferred to paper. Once the water has evaporated, the dyes remain on the sheet of paper. b) The pigments used in photosynthesis are not dissolved, but are present as solids embedded in membranes and proteins.

Polymerase Enzyme that catalyzes the construction of nucleic acids, the genetic material in the form of DNA and RNA. During cell division, the genetic material must also be duplicated. This is accomplished by the polymerases. But all viruses also need these enzymes to replicate their genetic material. Polymerases can be thought of as copying machines.

Porphyrin Large ring systems made up of 20 carbon and 4 nitrogen atoms, which serve as the basic building blocks of numerous biochemical dyes or pigments. Examples: Chlorophyll a and b, Hemoglobin (red blood pigment).

ppb Parts per billion $=1$ billionth part. 1 particle among 1 billion other particles. Equivalent to $1/10,000,000$th percent.

ppm Parts per million $=1$ millionth part. 1 particle among 1 million other particles. Equivalent to $1/10,000$th percent.

Precious metal Unreactive metals that occur unbound in nature and do not react with atmospheric oxygen or other substances. Examples: Gold, Silver, Platinum, Palladium.

Protein A substance that is made up of hundreds to thousands of amino acids.

Proton One of the three main elementary particles. It carries a positive charge, which is defined as +1. The actual charge is $+1.6 \times 10^{-19}$ C. Its absolute mass is very small, namely 1.66×10^{-24} g. The relative mass compared to the carbon atom is given as 1 u. Protons (and neutrons) contribute the lion's share to the weight of an atom. Symbol: p^+.

Protoporphyrin IX An important precursor for the natural production of chlorophyll or hemoglobin. Also referred to as Ooporphyrin.

Pteridin An aromatic double ring made up of 6 carbon atoms, 4 nitrogen atoms, and 4 hydrogen atoms. An important building block for folic acid (Vitamin B9) and riboflavin (Vitamin B2).

PVC Polyvinyl chloride, a plastic made from vinyl chloride units. Numerous uses include floor coverings, pipes, cable sheaths, dowels, and vinyl records. World annual production (2021): 37 million tons, polyethylene: 391 million t, polypropylene: 44.8 million t.

Quencher/Quenching Quenchers are substances capable of extinguishing fluorescence. In this process, the dye that fluoresces is not destroyed. As soon as the quencher is removed, the intensity of the fluorescence increases again. The effect of quenchers is based, among other things, on energy absorption or a "packaging" of the dye.

Radiant Energy Energy that is transported by electromagnetic waves. Examples: light, heat radiation (infrared light), UV radiation, X-ray radiation.

Radiationless Vibrational Relaxation If a molecule is excited with UV light, it is catapulted into a higher energy level by absorbing energy. In this excited state, the molecule oscillates back and forth and can collide with other molecules. In doing so, it transfers its energy to the collided molecule—like two billiard balls colliding. This process is referred to as vibrational relaxation. Relaxation means relaxation, slackening, in other words, the gradual transition from a highly excited state to a less excited state. Since no radiation is produced in this process, it is referred to as a radiationless process.

RCP Representative Concentration Pathway. In the Fifth Assessment Report of the Intergovernmental Panel on Climate Change (IPCC), four future scenarios are identified, which are designated as RCP2.6, RCP4.5, RCP6.0, and RCP8.5 according to the assumed range of the Earth's energy balance through solar radiation, given in watts per square meter, in the year 2100 (e.g., 2.6 W/m^2).

Reaction center Central location in an enzyme or during photosynthesis where the crucial biochemical reactions take place.

Receptor A protein or a group of proteins anchored in the cell membrane to which signaling substances can dock. According to the "key-lock principle", each receptor has a matching substance. Examples: a) Viruses infect their specific host cells using their surface proteins, which I like to compare with hands. If the receptor and the virus shake hands, the virus can penetrate the cell. b) The neurotransmitter (nerve messenger substance) acetylcholine binds to the nicotinic acetylcholine receptor in muscle cells at the motor end plate and opens an ion channel for sodium ions, without which muscles cannot move.

Red Algae Most of the approximately 7000 species are marine dwellers and macroscopic plants, i.e., such algae that can be seen with the naked eye. They perform photosynthesis using chlorophyll *a*, carotenoids, and the violet-red pigments phycoerythrin and phycocyanin, which particularly absorb green and yellow light. Edible red algae are cultivated in huge quantities and come to the market dried, roasted, or seasoned as "Nori" for sushi rolls.

Reflection In a uniform medium, the law of reflection applies: angle of incidence = angle of reflection or reflection angle ($\alpha = \alpha'$).

Reflectivity Ratio between reflected and incident light or sound waves. Example: The scattered reflection of light on rough (non-mirroring) surfaces. In terms of climate, reflectivity corresponds to the albedo effect.

Refraction (Snellius's Law of Refraction) When light passes through two different density media, such as air and glass or air and water, the light rays are refracted at the interfaces. When the light ray transitions from the optically thinner to the optically denser medium, the light is refracted towards the incident ray, i.e., the angle of refraction is smaller than the angle of incidence. When the light ray transitions from the optically denser to the optically thinner medium, the light is refracted away from the incident ray, i.e., the angle of refraction is larger than the angle of incidence. Formula: $n_1 \sin \alpha = n_2 \sin \beta$. This law was named after Willebrord van Roijen Snell (1580–1626).

Refractive Index Also referred to as index of refraction. The refractive index is a measure of the speed of light propagation in a specific medium. Air has a value of 1.0, water around 1.3, and glass about 1.5. The ratio of the refractive index of a medium to the refractive index of air determines the angle at which a light beam is refracted at the interface between air and the medium.

Rhizocarpic acid Yellow, fluorescent dye of the yellow-fruited sulfur lichen.

Rotifers Rotifers are multicellular microorganisms (zooplankton) with lengths from 40 µm to 3 mm and are also referred to as tube worms. The known 2000 species are almost exclusively found in freshwater and are very widespread, in every body of water, in every puddle. Rotifers consist of a head, a trunk, and a foot section. In water, they can swim in a rotating manner.

Russulumazin Natural fluorescent dye from the group of pteridines.

Scattering Unidirectional or diffuse reflection. All light rays are reflected by the material in very different directions "haphazardly". Almost all objects or particles scatter light and do not reflect it. The resulting color is white, like fog, clouds, or snow. Color is created solely by light absorption. If the particles are smaller than the wavelength of light, that is, smaller than 400–700 nm, a wavelength-dependent scattering, the so-called Rayleigh scattering, is the result. The gas molecules of the air are much smaller. Short-wave light, like blue light, is scattered much more strongly than long-wave light, like red light, since the scattering factor is inversely proportional to the wavelength, and in the fourth power (scattering $\sim \frac{1}{\lambda^4}$). Therefore, the sky appears blue (and the sunset red). If red light had a shorter wavelength than blue light, we would have a radiant red sky and blue sunsets (scary thought).

Sclerotin A structural protein that is primarily found in the outer skin of crustaceans and forms a hard shell with chitin.

Signal messenger substance Hormones. Chemical substances that transmit signals or information between organisms or between the cells of an organism in different ways.

Silicate Water-insoluble minerals or compounds made of silicon dioxide (SiO_2) or SiO_4 tetrahedral units, which can arrange themselves into chains, bands, and layers. The incorporation of other elements results in, among other things, desirable, valuable, and well-known silicates. Examples: Opal, Amethyst, Emerald, Aquamarine, Topaz, Olivine, Zircon, Talc, (Borosilicate) glass.

Silicon dioxide Quartz/Sand. The fundamental structural element of the various SiO_2 crystal structures is a tetrahedron, in which each silicon atom is surrounded by four oxygen atoms. The SiO_4 tetrahedra are connected to each other via their corners (oxygen) into a large crystal lattice. Formula: SiO_2.

Singlet Oxygen Excited oxygen molecule (O_2^*).

Singlet State All organic molecules, such as carbohydrates, fats, proteins, fragrances, etc., always contain an even number of electrons. This is because each atom-atom bond is mediated by two electrons, the so-called electron pair bond (in a double bond there are four, in a triple bond there are six electrons). The paired electrons of a bond possess an antiparallel (opposite) spin. The spin refers to the twist, the rotational movement, the angular momentum of an electron. This spin can (quantum mechanically speaking) take exactly two values. Put simply: left or right. If the rotational movement of the two electrons is opposite, this state is referred to as a singlet (S). This is because such molecules do not react with a magnetic field; they remain "singles". In contrast, in addition to the excited singlet states (S1 and S2), there are also excited molecules with parallel spin. These are in the so-called triplet state (T), because they interact with a magnetic field and split into three. This effect plays an important role in phosphorescence.

Starch A polysaccharide that consists of thousands of α-D-glucose units. Starch is present in all plants as an energy storage substance. It is not a uniform substance, but consists of 25% water-soluble amylose, a helix of about 200–5000 glucose molecules. The rest is the branched amylopectin, which is made up of several thousand glucose units.

Static friction Force that must be exerted to set a body in motion against its base.

Stretch vibration Vibration along the connection of two atoms, which arises from stretching and contracting. Human analogy: As if one stretches their arms upwards and then bends them again, like in a cheer. Also called valence vibration.

Structural protein Protein molecules that serve for the construction and stability of cells or organisms. They provide a solid "framework" (connective tissue). The sequence of their amino acids is regular and forms long fibers. Examples: Collagen, Keratin

Stinging thread Tentacle densely covered with stinging cells. Upon contact, poisonous barbed arrows shoot into the prey at an incredible speed of at least 10 m/s, paralyzing it.

Sugar kelp A large seaweed that is mainly found in the Atlantic, in the North and Baltic Seas, but also in the Mediterranean. Sugar kelp is the largest of all brown algae, carries out photosynthesis, is edible and tastes sweet due to its sugar substances. Hence its name. Sugar kelp can grow 1–4 m long and up to 30 cm wide and forms long wavy algae garlands.

Sulfur Dioxide Colorless, pungent-smelling, toxic gas. It is primarily produced by the combustion of sulfur-containing fossil fuels such as coal or petroleum products, which can contain up to 4% sulfur. Formula: SO_2.

Sulfuric Acid Colorless, oily, and hygroscopic liquid with strong corrosive properties. One of the most widely used substances worldwide. Its salts are called sulfates and hydrogen sulfates. Formula: H_2SO_4.

Sulfurous Acid Sulfurous acid (named Dihydrogen sulfite according to the IUPAC nomenclature) is a weak acid that is formed when sulfur dioxide is dissolved in water. Its salts are called sulfites and hydrogen sulfites. Formula: H_2SO_3.

Surfactant A washing-active substance consisting of two different molecular groups: The "head" is hydrophilic (water-loving) and the adjacent "tail" is lipophilic (fat-loving). Both properties are combined in one molecule, thus causing the dissolution of fats in water. In food, surfactants are referred to as emulsifiers.

Symbiosis Coexistence of living beings of different species for mutual benefit.

Tardigrades Tiny creatures with a size of 400–1500μm, which live in or on bodies of water and moist mosses. They look like little bears (although with 6 legs).

Tera 10^{12} = Factor 1 trillion. Symbol: T.

Tetrahedron Platonic solid (polyhedron) with four equilateral triangles as faces.

Thermal Equilibration A molecule excited by UV light can also release its energy as heat to the surroundings. This balance between excited and "de-excited" energy is referred to as equilibration (also equilibration).

Thermal radiation Energy transfer without material mediation. In the case of thermal radiation, there is no direct contact with a hot object (usually in contrast to heat flow and heat conduction). No heated matter (gas or liquid) is transported. All hot objects emit thermal radiation. In this way, energy also reaches the Earth from the Sun and can either be reflected on bright, shiny surfaces or absorbed on dark, matte surfaces. When absorbing thermal radiation, a body heats up and its temperature increases. Examples: In summer, people prefer to wear white or light clothing. Refrigerated vehicles are usually painted in light colors. Bags for hot chickens are lined with aluminum foil on the inside. Chocolate kisses are often packed in mirrored boxes. Black car paint gets much hotter in the sun than white paint. Rescue foils are coated in silver or gold.

Tipping point Critical threshold at which a small additional disturbance can lead to a guaranteed change on Earth. Examples: The ice sheets in Greenland and West Antarctica, the Atlantic circulation, the Amazon rainforest, or the coral reefs.

Total Reflection If the angle of incidence exceeds the critical angle, total reflection occurs, i.e., there are no more refracted light rays and the light is completely reflected at the boundary surface. The critical angle is defined as the angle of incidence that corresponds to a refraction angle of 90°. The critical angle only exists for the light path from the optically denser to the optically thinner medium.

Transferase Enzyme that catalyzes the transfer of a molecular group from one substance to another. Example: a) Hexokinase—it transfers a phosphate group to a glucose molecule, making it usable for our energy metabolism. b) Glycosyltransferase—it transfers a sugar molecule to a protein by binding to an OH group. This results in glycoproteins, which are important as structural components for cell-cell interactions and for mucus formation.

Transmission Passage of (light) rays through a medium without change in frequency.

Troposphere Lowest layer of the atmosphere up to about 12 km high. This is where the weather takes place.

UV Light UV light has a shorter wavelength than blue light and is divided into three categories. UV-A radiation ranges from 380 to 315 nm and UV-B light from 315 to 280 nm. UV-C radiation (218–100 nm) is even more energy-rich, but fortunately, it is completely absorbed by the atmosphere.

Virostatic Antiviral. Virostatics are medications that stop the replication of viruses.

Viruses In short: genetic material packaged in protein shells. Tiny organisms that consist only of a membrane and protein shells, inside which the viral genetic substance and possibly some viral enzymes are located. Viruses are not capable of life and reproduction on their own and therefore require host cells into which they penetrate. Size: 20–300 nm.

Vitamin C Ascorbic acid. A colorless, slightly sour-tasting solid. Ascorbic acid is enzymatically produced from glucose in plants. Therefore, it gets all its six carbon atoms from glucose and is thus a sugar derivative. It can easily give off one of its hydrogen atoms as an H^+ ion and has a pH value of 4–5 in aqueous solution, so it reacts quite acidic. In the body, vitamin C is involved in many biochemical reactions, for example in the biosynthesis of collagen/connective tissue. A deficiency of vitamin C leads to scurvy. Daily requirement: approx. 100 mg.

Vulpinic Acid Yellow, toxic dye that occurs in lichens and fluoresces under UV light.

Wash margin Deposits of plant and animal remains (or even garbage) that are washed up and left dry at the highest water level on the seashore.

Wavelength The distance between two successive wave peaks or troughs is called the wavelength of the wave. Examples: FM radio waves have a wavelength of about 3 m. Red light has a wavelength of about 650 nm = 0.00000065 m. Formula symbol: λ

Weight force Gravity. It causes weight on Earth and depends on the masses of the two attracting bodies and the distance between their centers. According to the law of gravity, the weight or gravitational force $F = m \cdot g$, where g is the acceleration due to gravity or the local factor. The weight force always acts in the direction of the center of the Earth and ensures that no person "falls off" the Earth, and that rain always trickles down.

Xanthophylls Oxygen-enriched, oxidized carotenoids. These also natural color pigments produced in plants contain, in addition to many carbon and hydrogen atoms, an OH group and/or an oxygen atom. The name comes from Greek and means "yellow leaf". Examples: Lutein (orange-yellow leaf pigment, also food colorant E161b), Capsanthin (red pepper pigment, also food colorant E160c), Zeaxanthin (yellow-orange leaf pigment, also in corn kernels, egg yolk, peaches, saffron), Astaxanthin (salmon, shrimp, lobster).

YAG phosphor Artificially produced compound with the chemical composition $Y_3Al_5O_{12}$. The abbreviation YAG stands for: Yttrium-Aluminum-Garnet, where garnet refers to a specific crystal structure. Cerium-doped YAG powder (YAG:Ce^{3+}) is used as a yellow phosphor in LEDs, which generate blue-violet light with the help of indium gallium nitride, thus converting it into warm white light.